The Self-Organizing Polity

About the Book and Author

Is the study of living systems a useful metaphor for political science? In this book, Dr. Dobuzinskis argues for further exploration of biopolitical models to explain the complexity of political theory and social change. His discussion emphasizes the new cybernetics, which considers not only self-regulating but also self-organizing or self-producing systems. Self-organizing systems operate in an autonomous sphere comparable to the autonomy of the political community and the political actors who compose this community. The autonomy of these systems is maintained through dynamic equilibration processes that entail not only the preservation of a given structure but also, at crucial times, the creative rearrangement of the existing structure and its transformation into a new pattern of relations. From this perspective, a political crisis is both a threat to the political system and the occasion of its renewal; stability may also mean decay.

Emphasizing the links that have developed historically between the natural and social sciences, this book is a reflection on the merits of and difficulties involved in representing the evolutionary process at the political level as the problematic reproduction of national communities and states.

Laurent Dobuzinskis is an assistant professor of political science at Simon Fraser University in Canada.

The Self-Organizing Polity

An Epistemological Analysis of Political Life

Laurent Dobuzinskis

Routledge
Taylor & Francis Group
LONDON AND NEW YORK

First published 1987 by Westview Press, Inc.

Published 2019 by Routledge
52 Vanderbilt Avenue, New York, NY 10017
2 Park Square, Milton Park, Abingdon, Oxon OX14 4RN

Routledge is an imprint of the Taylor & Francis Group, an informa business

Library of Congress Cataloging-in-Publication Data
Dobuzinskis, Laurent.
The self-organizing polity.
Bibliography: p.
Includes index.
1. Biopolitics. 2. Political science. I. Title.
TA80.D63 1987 320'.01'547 86-28261

ISBN 13: 978-0-367-29573-8 (hbk)

Contents

Acknowledgments

To those who have encouraged and facilitated what might have seemed my never-ending pursuit of this work, I offer my most sincere gratitude. In particular, I have benefitted immensely from the critical comments made by the members of my dissertation committee at York University as this book evolved from my doctoral dissertation: Professor H. Michael Stevenson, Professor David V.J. Bell, Professor Harold Kaplan and Professor Paul Medow. I would like also to thank the anonymous reviewer to whom a first draft of this book was submitted by Westview Press. The responsibility for any error or omission remains mine and only mine, of course.

I am grateful also to Barbara Barnett, who typed the last two versions through which this work has gone. But my most onerous debt is to my wife, Joan, who held the fort while I was "working on my book."

Laurent Dobuzinskis

The Self-Organizing Polity

Introduction

My purpose here is to evaluate critically both the methods and the subject matter of political science. Political scientists spend probably as much time writing about the methodology of political science as they do writing about politics. Some may think that this is unfortunate. But such an inclination is the inescapable consequence of the nature of our ill-defined subject matter. Methodological reflection is necessary for ensuring that what we write about is relevant to the concerns of those who are engaged in political activities and continually reproduce the political structures that sustain political life. This book is precisely an attempt to consider political life as a series of complex processes that the life sciences can help us to understand more fully.

Theoretical research of this kind must, of course, be informed by a discussion of concrete realities. The post-behavioralist movement has rightly denounced the tendency to postpone the analysis of emerging political issues until that time when they can be explained by means of a fully developed theoretical framework. Contemporary issues such as the crisis of the nation-state and the politics of man's relation with nature will be addressed. However, I argue that theoretical research ought not only to be practically relevant, but also to be epistemologically relevant, by which I mean that it should address the problems that pose a challenge to the dominant scientific paradigms.

Finally, there is a moral dimension inherent in any theoretical inquiry. I am concerned in particular with the reduction of reason to instrumental rationality which, in practice, takes the form of a confusion of the good with the imperatives of technology. Although the relationship between science, technology, and moral or religious values remains more complex than it is sometimes claimed to be by critics of our technocratic age, it is also true that the technological *Zeitgeist*, as it were, acts as a powerful leveler. As far as politics is concerned, one of the most far-reaching technological developments of the last two decades has been the tremendous progress of informational technology and its transformation into a kind of "decisional technology" that has had a decisive impact on public policy-making. It has contributed to the opinion of the technocratic elite that politics is a derivative, if not altogether dysfunctional, activity. My contention is, however, that in theoretical cybernetics we can find powerful arguments against the manipulative practices that we have come to associate with applied cybernetic techniques of decision-making in large bureaucracies. I do not intend to show that "pure" cybernetic theory has become the victim of unscrupulous technicians motivated only by the pursuit of selfish gains, for there are clear, logical connections between some theoretical principles in cybernetics and the technocratic ideal of a rationalized, centralized mode of societal control. But there are several distinct trends in theoretical cybernetics, some of which point towards a radical critique of both science and society.

Political scientists interested in systemic and cybernetic analogies have almost completely ignored the second generation of cybernetic models whose origins are traced hereafter. They are aimed at the formalization of the "problem of order"—the perennial problem of articulating and ranking the principles that give a meaning to the inner and outer realities of our lives. Of course, the problem of order has always been a concern for philosophy, but with the new cybernetics it has become a concern for science *qua* science.

The Old and the New Cybernetics

Twenty years after the publication of the first edition of Karl Deutsch's *The Nerves of Government*, it may sound trite to claim again that cybernetics can make a contribution to the scientific study of politics and to the practical application of scientific knowledge to policy-making. But I will argue that the second generation of cybernetic models pose new questions and suggest new answers. The first three chapters of this book provide an in-depth analysis of their substantive content but, to avoid any misunderstanding, a brief historical account of their development is in order here.

We can trace back the origin of systems theory to the 1930s when Ludwig von Bertalanffy began to look for a way out of the controversy between the mechanistic and vitalistic approaches in biology. The former did not offer any clue about the complex interactions out of which the unity of living things emerges; the latter was decidedly too mystical to qualify as a scientific conception. Bertalanffy quickly came to the realization that organisms should be viewed as "open systems" whose behavior could only be understood by expanding the laws of conventional physics and chemistry.[1] Although he had from the beginning envisaged the possibility of further generalizations of the organismic principles he was elaborating, he waited until the late 1940s to publish his findings. Since then General Systems Theory (GST) has developed out of his work.[2] This theory follows an "empirico-intuitive" approach whereby statements of a general nature are arrived at in the wake of the detailed examination of a variety of concrete systems.[3] The ultimate goal is to formulate mathematical expressions describing modes of transition from one system state to the other, whatever the system might be; Bertalanffy favored systems of differential equations for this purpose.[4] But the weaknesses of the inductive method, coupled with Bertalanffy's lack of mathematical sophistication, explain in part why the ambitious objectives ascribed to GST have not been met. The kind of problems raised by this school of thought, centered as they are on dynamic phenomena like finality, growth differentiation, the thermodynamics of life, etc., are nevertheless still worth investigating. However, this cannot be done without the help of cybernetic concepts and methods.

Cybernetics originated in the research carried out by Norbert Wiener in the 1940s. Wiener was the author of the seminal work *Cybernetics or Control and Communication in the Animal and the Machine* (1948), and benefitted from his collaboration with J.H. Bigelow and the neurophysiologist A. Rosenblueth. He was involved with the former in the design of servomechanisms—i.e., machines that can, in some respects, be regarded as "intelligent"—and his association with the latter suggested far-reaching applications of what he had discovered about such machines.[5] Feedback is, of course, the central concept of cybernetics. For the moment let us simply define feedback as a closed information loop. When information about previous system states is used to counterbalance transformations taking place within the system, a "negative" feedback is at work; when, on the contrary, information contributes to the acceleration of internal transformations, "positive" feedback is at work. The first type is characteristic of adaptive, stable behavior, like that of the human body with respect to temperature, or of a heating installation controlled by a thermostat. The second type causes exponential growth or regression; the recent and dramatic increase of the world population is a good illustration of positive feedback at work. N. Wiener and electronic engineers were particularly interested in negative feedback processes. Undeniably, control feedback is an essential dimension of cybernetics but the importance ascribed to that concept varies from one author to the other.

The technological applications of cybernetics, which also gained from the parallel development of information theory, have been almost exclusively concerned with control feedback. In this perspective systems maintenance is paramount. Moreover, first-generation cybernetic models typically incorporate optimizing algorithms intended to reduce the complexity of the systems under consideration to a very limited range of outputs. What is known in the management sciences as "systems analysis" is rather directly inspired from this kind of systems engineering.

However, cybernetics is not reducible to technology. The theories of cybernetic machines or automata developed by J. von Neumann, W.R. Ashby, and W. McCulloch[6] are pure scientific abstractions, even if they can be instructive in the design of concrete automated machines. In particular, when compared to N. Wiener's dazzling combination of mathematical sophistication and philosophical erudition, R.W. Ashby's work may appear more mundane. But his understanding of the significance of cybernetics for the study of living systems has turned out to be more illuminating. In particular, being less concerned than other cyberneticians of his generation with the technological applications of cybernetics, Ashby was in a position to move cybernetic theory out of the domain of *techne* and to ground it in the domain of *theoria*, especially in his contention that the materiality of the cybernetic machine was an irrelevant consideration.[7] His theory of machines is a theory in the fullest sense of the term, and not a treatise on machine building. On the basis of this argument, G. Günther suggests that cybernetics rests on a subjective ontology; its world is populated by autonomous subjects capable of making unpredictable choices.[8] Actually this remark more accurately applies to the "new cybernetics," in the development of which Günther has played an active part, than to the first generation of cybernetic models. With hindsight, Ashby appears to have been a transitional figure between the old and the new cybernetics.

It is perhaps too simple to suggest that the difference between the old and the new cybernetics lies in the importance that they attribute to the concept of positive feedback—that is, almost nil in the case of the former, paramount in the case of the latter. But it is true that the rediscovery, so to speak, of positive feedback in the early 1960s laid the groundwork for the formulation of a dialectic of feedback that finds its most articulate expression in the theory of self-organization. Three symposia on the subject of self-organization were held in 1959, 1960, and 1961.[9] During the mid-1960s the initial impetus weakened, perhaps because of the lack of direct application. H. von Foerster and the other members of the Biological Computer Laboratory of the University of Illinois still kept the movement alive, and it resurfaced in the 1970s, partly as a result of the deadly blow dealt by Ilia Prigogine to the universality of the second law of thermodynamics. (The non-equilibrium systems upon which Prigogine carried out his experiments possess the capacity to organize their rate of entropy production and thereby to alter their own structures.) The idea of self-organization has since been popularized by the writings of Erich Jantsch and Edgar Morin.[10] Chapter 3 of this book is devoted to an exposition of the central themes of this approach; but, at this point, it is sufficient to characterize the new cybernetics as an attempt to look beyond the regulation of the structures of a given system by raising the question of their origin. And that origin is found in the *autonomy*—in the literal and etymological sense of the term—of organizationally closed complex systems capable of reacting upon their own structures.

The New Cybernetics and the Science of Politics

The theory of self-organization offers a formalized, albeit somewhat incomplete, description of the processes through which complex systems, feeding on both entropy (or disorder) and "negentropy" (or order), generate their *own* structures. Thus one can speak of the "creative function of disorder." Such a conception suggests original points of view for the discussion of three current problems in political theory, including both political philosophy and empirical theory. First, what could be the relevance of the concept of system in the post-behavioral, post-positivist age which political theory has now entered. Second, on what basis could one establish the complementarity of traditional political theory and empirical theory, that such an epistemological context presupposes. Finally, what should be the methodological and practical implications of the contemporary rediscovery of the naturalness of man-the-political-animal, prompted by the ecological crisis. This book does not propose definitive answers to these momentous questions; rather it argues that i) a shift of emphasis in political inquiry, from a preoccupation with explaining how political actors act as they do in a given societal context, to a creative exploration of what is thereby learned and achieved that contributes to the production and reproduction of our individual and collective identities; and ii) the adoption of "dialogical" and diachronic methods of analysis, i.e., methods based on a multi-valued logic and explicitly incorporating a temporal dimension, are two prerequisites for a fruitful synthesis of normative and empirical political theory and for the furthering of the already lively dialogue between natural and social scientists.

The emphasis placed by the second generation of cybernetic models on the reflexive character of systemic/cybernetic concepts and methods reveals, in contrast, the epistemological paradox inherent in David Easton and Karl Deutsch's attempts to ground their own positivist behavioral approaches in what turns out to be a non-positivist paradigm.[11] Admittedly, this sounds like a plan for flogging a dead horse, as systems theory has been criticized *ad nauseam* in political science, found wanting in many respects, and is no longer considered relevant by most researchers, even if textbooks still rely on it. However, it is imperative to rescue some of the assumptions of systems theory and of cybernetics from this debacle for they give a substantive content to the notion of organized complexity.

The new cybernetics also suggests a way out of the longstanding quarrel between the ancients and the moderns; the controversy opposing empiricists and philosophically minded political theorists has been rather inconclusively settled in favor of the latter, to the extent that it is generally realized now that the behavioralists' rejection of metaphysics was itself based on a particular metaphysical view which has been losing ground in the philosophy of science since the beginning of this century. Yet we still lack a well articulated *problématique* to support on-going research. In the wake of the so-called post-behavioral revolution, a multitude of ad hoc philosophical perspectives and corresponding methods of inquiry have been suggested with little success. But the theory of self-organization provides us with an opportunity to pose anew, and in terms that are relevant to current concerns, the perennially fundamental problem: What is human nature? I do not propose a final answer to that problem, but the new cybernetics suggests that man, being the most complex and adaptive living system, is a fully autonomous subject. In unpredictable ways, man actively contributes to his reproduction and also to that of his natural and social environments, or the biosphere and the "noosphere." The biosphere is itself conceived as a complex field of interactions among self-organizing systems; and the noosphere, by which I mean the sphere of communicative action, including language and culture, also follows a developmental logic of its own. The implications of this perspective for political inquiry are varied and challenging.

Furthermore, a neo-cybernetic approach to politics suggests a series of interesting linkages between the natural and social realms. To reintroduce the notion of human nature poses the question of knowing what is human in nature and what is natural in man. Indeed, as Serge Moscovici puts it, the "natural question" is the dominant problem of the 20th century.[12] Here again, the theory of self-organization opens up the possibility of a reappropriation of older philosophical views, including Aristotle's *physis*. Of course, it would be bad science as well as bad philosophy to simply return to a pre-Galilean idea of nature. But cybernetics and several theoretical developments in thermodynamics and biology invite us to look critically at the dogmatic separation between nature and culture which is still characteristic of the social-scientific world-view. Is there really a gap between physico-biological and socio-cultural realities? When we look at our natural environment as a field of interactions among self-organizing systems, then man's social mode of existence becomes a measure of his naturalness. If so, could our knowledge of self-organizing natural systems inform our knowledge of society and vice versa? Far-reaching as these questions may be, they have been raised and debated over the last ten years by a loose group of French natural and social scientists, among whom Edgar Morin is perhaps the most influential. Morin's fascinating, if sometimes less than rigorous, elaborations upon the theory of self-organization provide useful bearings throughout much of this book.

The concepts of production and reproduction, commonly used in biology and economics, have less commonly been applied to the analysis of political life.[13] But if we are to understand politics as a creative process through which individuals and groups seek to achieve evolutionary potentials, we need to study political activities in relation to the production, destruction, and regeneration of the various dimensions of societal order.

The notion of evolutionary potential is offered here as an alternative to the familiar but pointless antithesis between determinism and randomness. Change is a problematic concept that can be used only within the context of an axiomatic construction of reality; it will be argued (see Chapter 4) that the construction which is most relevant to political inquiry consists in an evolutionary paradigm for it gives meaning to the belief tacitly shared by political actors in the continuity of the constantly changing political community in which they live.

A recurrent theme throughout this book is that the organization of evolutionary potentials into actual structures takes place along three dimensions: experience, consciousness, and self-consciousness. While all these dimensions refer to the fundamental autonomy of political actors, the first one is the least concerned with their reflective, subjective attributes. Consciousness, as this term will be used hereafter, extends from subconscious desires, insofar as they are actuated in political demands, to what Anthony Giddens calls "practical consciousness," i.e., a "non-discursive, but not unconscious, knowledge of social institutions—as involved in social reproduction."[14] On the other hand, self-consciousness develops through discursive relations among fully cognizant subjects.

Self-regulating processes of structure maintenance will be analyzed mostly along the first dimension. Processes of identity formation in a political community, above and beyond the ever changing conditions of that community's existence, will be discussed along the second and third dimensions. But to do so, we must reconsider the level-of-analysis problem which is usually presented as that of a choice between micro and macro-analysis; micro-analysis focuses on individuals and small groups, whereas macro-analysis is concerned with the performance of the political system defined as a subsystem of society. I shall argue that macro-analysis is itself a multi-level approach. The primary level consists in the regulatory functions of the political system, that is to say the co-ordinating and coercive practices by which a structurally differentiated society maintains its integration.[15] But, looking beyond the institutional and behavioral boundaries of the political system, we ought also to raise the problem of the origins of the norms and values embodied in the political system, for a societal community reproduces itself by evolving new values through a global political process that transcends analytical distinctions among functionally specialized subsystems.[16] The generation of norms and values poses a metatheoretical problem whose solution requires the formulation of a reflexive strategy of cognition. This is because the dominant themes of political theory contribute to the self-production of the polity.[17]

In brief, this book proposes to look at politics as an evolutionary process. Political life, from this perspective, involves a subtle dialectic of invariance and change. The recursive interaction of autonomous political actors and sub-groups with the political structures of the political community to which they belong delineate a field within which living systems of a new order (such as bureaucracies, nation-states, etc.) emerge from time to time, develop and evolve into still newer forms. The logic of this process does not depend on any subjective characteristic. However, the unique form assumed by these political beings reflects the practical and reflexive consciousness of the subjects who produce and reproduce them. Consequently, in order to understand self-organizing processes in political life, one must also formulate a strategy for relating empirical concerns to normative reflections.

Now the realization of a research program of this kind depends on the development of appropriate methods. Two fundamental questions arise in this respect.

First, what are the methodological implications of a position that combines a naturalistic vision of man-the-political-animal with a humanized vision of nature? The answer does not lie in the straightforward application of biological or ethological knowledge to the explanation of political life, but rather in the recursive exploration of organized complexity along a paradigmatic "loop" linking physics to biology and to political anthropology. My discussion of the notion of societal evolution will illustrate the merits of this approach (see Chapter 4). Second, what are the implications of an approach which re-introduces the notion of "self" in political analysis? I shall argue that the adoption of a cybernetic paradigm must be preceded by, and articulated to a discussion of the various dimensions of consciousness.

Third, I shall argue that we cannot explain political phenomena, but we can nevertheless construct models of the political process to simulate alternative societal priorities. To model complex systems one must first dig out, so to speak, most of the potential contradictions present in the system in order to render explicit the various possible developmental paths. This requires, as already stated, the adoption of a multi-valued logic and a diachronic methodology.[18]

Synopsis

This book is divided into eight chapters which are briefly summarized hereafter.

Chapters 1 to 3 are devoted to the analysis of the foundations of systems theory and cybernetics; they follow a logical progression from a discussion of the complexity inherent in the concept of system itself (chapter 1) to that of the complexity of the self-regulated systems described by the first generation of cybernetic models (chapter 2) and to a reflection on self-organizing complexity (chapter 3). Chapter 4 broadens the perspective presented in chapter 3 by applying it to the analysis of evolutionary process in both natural and societal systems. Chapters 5 to 8 focus on the more specifically political aspects of societal evolution. Chapter 5 proposes a three-dimensional grid which is articulated around the ontological notions of existence, consciousness and self-consciousness. It is then used for the reconstruction of the subject matter of political science in the next three chapters. Chapter 6 deals with the regulatory function of political institutions. In chapter 7, the evolutionary paradigm provides a starting point for a discussion of the problem of the re-production of national communities organized as political systems by autonomous actors entering into cooperative and conflictual relations in a public domain. Finally, chapter 8 suggests agenda of theoretical research on the interdependence of the social and natural sciences envisaged from the standpoint of a subjectivist epistemology.

Notes

1. See L. von Bertalanffy, "General Systems Theory—A Review," in W. Buckley (ed.), *Modern Systems Research for the Behavioural Scientist* (Chicago: Aldine, 1968), p. 13. Note that, at the same time, W. Kohler had expounded similar ideas; see W. Kohler, "Closed and Open Systems," in F.E. Emery (ed.), *Systems Thinking* (Penguin Books, 1969), pp. 59–69.

2. There is a General Systems Society which publishes the *General Systems Yearbook*.

3. L. von Bertalanffy, "General Systems Theory," p. 13.

4. Bertalanffy is the first one to admit that such a tool is of limited value, see ibid., p. 14, and his *General Systems Theory* (New York: Braziller, 1973, 2nd ed.) pp. 56–77.

5. See J. de Rosnay, *The Macroscope* (New York: Harper & Row, 1979), Chapter II.

6. See J. von Neumann, *Theory of Self-Reproducing Automata* (Urbana, Ill.: University of Illinois Press, 1966); W.R. Ashby, *An Introduction to Cybernetics* (New York: Wiley, 1956).

7. See ibid., p. 1.

8. See G. Günther, "Cybernetic Ontology and Transjunctional Operations," in M.C. Yovits et al. (eds.), *Self-Organizing Systems* (Washington, D.C.: Spartan Books, 1961), pp. 313–92.

9. See M.C. Yovits and S. Cameron (eds.), *Self-Organizing Systems* (New York: Pergamon, 1960); M.C. Yovits, J. Jacobi, and J. Goldstein (eds.), *Self-Organizing*

Systems (Washington, D.C.: Spartan Books, 1961); H. von Foerster and G.W. Zopf (eds.), *Principles of Self-Organization* (New York: Pergamon, 1962).

10. E. Jantsch, *The Self-Organizing Universe* (Oxford: Pergamon Press, 1980), and E. Morin, *La méthode*, vol. 1 (Paris: Seuil, 1977).

11. For an elaboration of this point see Chapters 3 and 6.

12. See S. Moscovici, *The Human History of Nature* (New York: Harper & Row, 1977).

13. Admittedly, the problem of the reproduction of the state has become a dominant concern for neo-Marxist political economists (see Chapters 6 and 7 *infra*); on the relationship between biological and political reproduction, see Mary O'Brien, "The Politics of Reproduction," unpublished dissertation (Toronto: York University, 1976).

14. Anthony Giddens, *Central Problems in Social Theory* (London: Macmillan, 1979), p. 24.

15. See Chapter 5.

16. See Chapter 6.

17. See Chapter 7.

18. See Chapter 8.

1

The Complexity of Complex Systems

To give an account of evolutionary processes in nature or in society without defining a variety of interrelations as systems (e.g., the eco-system, the economic system, etc.) is virtually impossible. The organization of a system is typically complex. In this chapter, I discuss the complexity inherent in the very notion of systemic organization. Chapters 2 and 3 deal with the dynamics of complex systems (i.e., self-regulation and self-organization). The rationale for this exercise is that political life is assumed here to be a prime example of organized complexity. Unfortunately, many of the traditional concepts used by the students of politics are too unrefined to express the puzzling and manifold implications of complexity. While some political scientists may feel that the techniques described hereafter are not immediately relevant to their concerns, I suggest that they are essential for understanding the logic of a wide variety of natural and societal processes, including many aspects of political life.

Political scientists and sociologists are all familiar with the term "system." When S. Chodak characterizes social change as a process involving "a growing societal systemness"[1]—by which he means differentiation within greater independence—he needs to add little to that phrase by way of definition. Yet this widespread usage is also the source of serious misunderstandings. In the first place, the term "system" is not always employed in the context of a theory where it assumes an unambiguous meaning. Considering that the term itself antedates systems theory, this is not surprising but remains nonetheless confusing. Moreover, in political science, systems theory as it was originally formulated by David Easton is in many (and too many) respects a truncated—although not worthless—application of the systemic paradigm.

Under these conditions, it seems necessary to begin a reformulation of the systemic approach and of its application to societal analysis by critically reviewing the various definitions of the concept of "system" found in the literature of the natural and social science. However, a *caveat* is a propos here. An ideal terminology should be evaluationally neutral. Thus it is not uncommon to find excellent definitions of important systemic concepts proposed by authors whose premises are quite different from mine; in such cases I do not hesitate to adopt as mine the definition in question, taking care to stress the "proper" context of reference.

Definition of the Concepts of System and Complexity

The concepts of system and complexity are very closely related. Indeed, the chief merit of the systemic approach is that it constitutes an attempt to understand complex problems as belonging to a specific class of scientific problems. Now, complexity can be defined in several ways. For Edgar Morin:

> The problem of complexity is that of phenomena which cannot be reduced to the simple thought patterns of the observer. In other words, complexity will first manifest itself, for the observer, in the form of obscurity, doubt, ambiguity, or even paradox or contradiction.[2]

Simple phenomena and simple thoughts are clear and unambiguous, even if they may be complicated, that is to say hard, but not impossible to untangle. Not only does complexity imply extreme complicatedness, but also irreducible uncertainty.

A similar argument is put forward by N. Georgescu-Roegen. According to the latter, there are scientific problems not amenable to formal logic.[3] Reinterpreting Georgescu-Roegen's argument, it can be said that reality is both continuous, as far as the *temporally* invariant properties of the elementary components of both nature and society are concerned (e.g., D.N.A., or the idea of God), and discontinuous with respect to the *dynamic* arrangements of these elements. Formal logic, on the contrary, introduces discrete categories frozen in time. It creates discontinuity between the classes where there is in fact continuity, as between the biological and social realms, which are characterized by similar organizing principles,[4] and creates continuity between the elements of each class where there is in fact discontinuity, as between a cell and man. Yet no symmetrical rearrangement of the classes could solve that dilemma because the discontinuity between, say, a cell and man within the class of living beings is due to the passage of time and not to the intervention of an alternative logical criterion. Complexity comes from the existence of what Georgescu-Roegen calls a zone of "penumbra" between the concept and its opposite;[5] it is in this interstice that discontinuous and irreversible qualitative changes take place. For instance, there is no clear-cut distinction between living and non-living matter, since in between we find puzzling forms such as crystal-viruses. Hence, a logic of complexity is necessarily dialectical.

W. Weaver's analysis of the implications of complexity for science brings us closer to a rigorous definition. He distinguishes between three problem situations: simplicity, unorganized complexity and organized complexity.[6] Simple problems like the two body problems of classical physics involve a manageable number of variables, and yielded to mathematical analysis in the 19th century. A large number of randomly interacting entities, which nevertheless display orderly properties, give rise to problems of unorganized complexity; these can be solved through the techniques of probability calculus. Voting behavior is a good example of this. Problems of organized complexity, on the other hand, "are problems which involve dealing simultaneously with a *sizeable number of factors which are interrelated into an organic whole*."[7] These problems, according to Weaver, cannot be properly handled through probability analysis and raise very serious methodological questions.

Besides, the "sizeable number of factors" composing an organic whole are likely to be interrelated through conflicting as well as complementary processes. In other words, the state of equilibrium of a complex organic whole lies somewhere in the penumbra between the concept of stability and its opposite. The term "equilibrium" is indeed inadequate; it is preferable to think of the behavior of complex organisms in terms of "self-balancing fluctuations."[8] Because, at any moment, the resulting effect of environmental and internal conditions upon the complex organism remains uncertain, there is the possibility that the latter will be free to chose a radically unpredictable course of action affecting, in return, its environmental and/or its internal parts. This we call regulation; to state, however, that the regulatory activities of a complex organismic whole potentially entail free acts does not necessarily mean that such is always the case.

Up to now we have been discussing complexity in abstract terms. (Incidentally, the notion of organic whole was also left intentionally undefined.) There are several levels of organized complexity corresponding to various organizational requirements. Edgar Morin suggests that an attentive look at the complexity of living systems, though not sufficient to account for the "hypercomplexity" of the human phenomenon, allows us to further explore that concept.[9] The rationale for borrowing analytic tools from biology is further argued below,[10] but it seems *a priori* justified to search for available models of complex processes where we find them which is in biology. This is only a methodological shortcut and political scientists should forge their own tools. At this moment, though, too little has been done in that respect.

E. Morin distinguishes between four levels of complexity to which correspond four types of systems, namely "systemic," "open" systems, "cybernetic," and "self-organizing" complexity.[11] To discuss each one of them is an opportunity to define the concept of system in the most complete manner.

Of Wholes and Parts

At the highest level of generality, a system is a set of non-randomly interacting elements such that the whole is more than the sum of its parts.[12] In physics, the atom is an example (an etymological paradox!)—its irreducible character evidenced by the reaction of nuclear fission. And so is the organism in biology but, in that case, the irreducible property of the whole, namely life, is more elusive. The downfall of Driesch's vitalist dogma does not constitute a refutation of the validity of the concept of system in biology but it compels us to eschew categorical interpretations of the "systemness" of complex entities. At this point it is still premature to propose a definition of the political system; it suffices to say that politics is not comparable to some kind of competitive market where individual decisions combine resulting in "optimal" political choices. Power relations and other aspects of political life are structured by rules that, together, are constitutive of interacting institutional and cultural systems. So far all examples have been, at least partially, of a physical nature but examples of conceptual systems also abound, such as the structure of group in mathematics (set theory). (The problem of the reality of the existence of concrete systems in the social sphere is discussed later; let us simply suggest that the distinction between concrete and conceptual systems is somehow artificial insofar as the subjective and the objective often overlap.)

The systemic approach constitutes a logic of wholes called for by A. Angyal more than three decades ago.[13] L. von Bertalanffy's pioneering work rests squarely on the assertion that the Aristotelian holistic and teleological world view, displaced by the scientific revolution of the 16th century, should regain its lost preeminence and that systems theory makes it possible.[14] Such a perspective allows us to reestablish the "aristocracy of form" that, according to Hans Jonas, Newtonian physics had replaced with the "democracy of matter" because, in the case of the latter, "No longer something original in its own right, form is the current compromise among the basic actions of aggregate matter."[15]

To define wholeness as form though does not go without ambiguities, insofar as "The word 'form' has many meanings, such as shape, configuration, structure, pattern, organization, and system of relationships."[16] In physics (e.g., quantum physics) and in biology (e.g., embryology) holistic notions of spatial symmetry and configuration do play a limited role in theoretical models. Although most scientists still shy away from problems such as the essential relationship between the concept of organization and geometrical principles, there seems to be a growing realization of their importance.[17] Hence the relevance of topological research in mathematics and especially the pioneering work of R. Thom for the study of the transformation of complex systems.[18]

The social space is more difficult to comprehend insofar as it involves relationships between symbols and consequently is perceived at a mediated level of consciousness. Yet, I posit that it is possible to identify overarching morphogenetic principles applicable to both the natural and the social spaces. These, in turn, will suggest specific hypotheses for the study of, for instance, embryological development, language, economic growth, or political life.[19] Unfortunately, many institutional and attitudinal obstacles stand in the way of the kind of interdisciplinary research required for that purpose.

In a holistic perspective, the discontinuous, singular, or "catastrophic" (i.e., due to sudden qualitative changes) properties of the system under investigation no longer appear to be the exception to an underlying regularity revealed by an analysis of its parts. No longer can we think in terms of irrelevant superstructures as opposed to a fundamental infrastructure, or ephemeral forms as opposed to an immutable content.[20] For Jacques Sauvan, traditional science, with its emphasis on experimental verification and its refusal to consider singular events as significant, has succumbed to the temptation of the myth of the eternal return, at the cost of becoming incapable of understanding human experience.[21] As Edgar Morin argues: "While the two concepts of event and system are usually dissociated, we must, on the contrary, try to conceive how they are indissolubly linked."[22]

Physics and biology would benefit from such a synthesis but the social sciences even more so. Their respective subject matters are replete with singular events and unexpected transformations from a unique mode of socio-cultural organization to another. In that respect, the statisticians' insistence upon the need to develop time series data should be taken with a grain of salt. Has the social system considered remained unchanged in its wholeness during selected time period? Probably not if it is an extended one. "Like the aristocratic nobility of the 18th and 19th centuries [sophisticated time series data], are beautifully refined—but increasingly irrelevant."[23] One-shot studies and isolated data, conversely, should not be discarded and have their place in a systemic model of social change.

Despite its merits a holistic approach—as opposed to an analytic reductionist one—presents some serious difficulties. If complex organic wholes do exist, then what is the nature of a social system? and what are the ethical implications of whatever answer we decide to give to these questions?

Too often those who take part in this debate seem to poke at a strawman of their own creation. The two extreme positions consist in stating that either wholes do exist and society constitutes a super-individual, or wholes do not exist as distinct from the sum of their parts and thus individuals' actions—motivated by generally rational criteria—are the sole cause of social phenomena (i.e., "methodological individualism").

What some authors deny, however, is the existence of a collective interest superior to and distinct from individuals' interests, and only that. Karl Popper, the most severe critic of naive holism, who "as a rationalist" wishes and hopes for a reduction, admits that "there may be no reduction possible; it is conceivable that life is an emergent property of physical bodies."[24] He even goes as far as to make the following plea: "let us work in every case the arguments for the emergence in detail, at any rate before attempting reduction."[25] Joseph Agassi argues that Popperian individualism is not incompatible with what he calls "institutionalism-cum-individualism."[26] In such a perspective, society represents the conventional means of co-ordination between individual actions. Agassi acknowledges the possibility—even the necessity—of analyzing institutions as such, independently from the study of the psychology of their members, but he rejects categorically the idea that collective organizations have aims and interests of their own. I would like to argue that such a view is unnecessarily restrictive.

Certainly, too many people, too many times have been saved from themselves against their own will by prophets of an illusory collective destiny. However, these totalitarian experiments owe much to the very conception of the individual, inherent in liberalism, as a self-contained unit of action independent from, if not set against, his fellow human beings. Totalitarianism is rendered possible by the atomization of the political community, by the substitution of the notion of a shared humanity by that of the individual, embodying what P. McCormick has called an "unproblematic self."[27] The early liberal thinkers were convinced that, while an individual enjoys immediate and unrestrained access to his innerself, his feelings and emotions are entirely private and uncommunicable. According to Robert M. Unger, liberal psychology radically separates facts from value and desire from reason.[28] Thereby it contradicts the necessary condition for the existence of a political community, inasmuch as such a community entails the search for a common good and the will to pursue it. Hence the totalitarian temptation of recreating the organic unity of the polity by the coercive manipulation of all the aspects of life in society for the purpose of confusing facts with values and history with ideology. The state or the party then becomes a mythical totality that promises to reconcile alienated individuals with the historical nation or "progressive" humanity.

An attempt to return to a pre-liberal understanding of the relationship between human nature and political life would be futile. But the kind of systems approach advocated here offers an opportunity to reconsider this fundamental relation. Organized complexity transcends the traditional distinction between parts and wholes. To state that the polity is more than the sum of the individual political actors it includes is, at best, an approximation. The component parts of any societal system are not individual as such, or even societal roles, but the outcomes of a multiplicity of self-organizing interactions among autonomous subjects who retain their ability to make unpredictable decisions (including that of entering into countervailing interactions). The wholeness of the parts, so to speak, is indeed a condition of the generation of the variety required for the restructuration of a self-organizing system, as it will be shown. In other words, to point the existence of autonomous societal systems in no way implies that we surrender our freedom to a collective entity. Thus I can see no reason why one could not share the ethical concerns characteristic of methodological individualism without subscribing to its methodological premises. Therefore, with A. Etzioni, I posit that:

> A social collectivity has a self, not as a reified or metaphysical category, but as the carrier of an active orientation. Collectivities can set goals, commit themselves to their realization and pursue them.[29]

The Complexity of Open Systems

Having established that a system is a complex organized unity, it remains to show that such a unity is not realized in isolation, but rather through structural coupling between the system and its environment. As E. Morin suggests:

> Another order of complexity is reached when the system is 'open,' that is, when its existence and the maintenance of its diversity are inseparable from interrelations with the environment, interrelations whereby the system draws on the outside world for material/energy and, at a higher level of complexity, elicits information. Here we have a truly complex and ambiguous relation between the open system and its environment, vis-a-vis which it is at the same time autonomous and dependent.[30]

A system is said to be closed when it is, for all practical purposes, isolated from its environment. The distinction between open and closed systems always entails a certain degree of arbitrariness. However, it is clear that, for instance, non-organic chemistry is all about phenomena occurring within precisely defined boundaries beyond which the *ceteris paribus* rule prevails. On the contrary, life, under all its forms, cannot be sustained in isolation. Living systems are open systems *par excellence*. It is meaningless to talk of the metabolism of a mineral, but is just as meaningless to look for instances of static equilibrium in the living order since open systems are exposed to external disturbances that affect their internal states to a greater or lesser extent. Similarly, "We may [view] political life as a system of behavior imbedded in an environment to the influence of which the political system itself is exposed and in turn reacts."[31]

The Thermodynamics of Open Systems

Ilya Prigogine and Isabelle Stengers trace the origin of a science of complexity and of the concomitant evolutionary paradigm to the discovery of entropy.[32] According to the First Law of Thermodynamics, the overall quantity of energy in the universe remains constant. But the Second Law introduces the notion of irreversible change; it posits that the quantity of usable energy decreases over time and the quantity of dissipated heat increases—heat cannot flow back into the fuel that an engine burns! One way of defining entropy is precisely to equate the dissipation of heat with entropy production. A more sophisticated definition refers to stochastic processes and equates entropy with disorder. For example, the molecules of a liquid or gas in a container where heat is evenly distributed and where, therefore, entropy is maximum, behave totally randomly. It would be simplistic to propose that the fundamental scientific problem of the day is that of reconciling the universal tendency toward entropic decay with the ever-increasing complexity of living systems. But it is a good starting point. Complex organized open systems produce negative entropy, or "negentropy."

Were it not for the Second Law of Thermodynamics, the distinction between closed and open systems would remain a tautology. L. von Bertalanffy insists that the law applies only to closed systems: in these, and in only these, entropy increases ineluctably until an equilibrium of sorts is reached, i.e., total randomness. In N. Wiener's terms, this is "nature's statistical tendency to disorder."[33] Incidentally, one might hope that those social scientists who are so concerned with the equilibrium of social systems would take into account the implications of the notion of thermodynamic equilibrium.[34] Equilibrium can mean disorder, ultimate chaos; organization or order, on the contrary, presupposes some degree of disequilibrium. (These ideas are developed further in Chapter 3.)

To state that universal entropic decay does not affect open systems is not absolutely correct. As it was already pointed out, the very notion of open system is somewhat vague. Besides, negentropic states are always achieved at the expense of the environment. It remains that open (living) systems (e.g., the human organism) survive, at least temporarily, without experiencing the wear and tear of mechanical systems. And from mutation to mutation, life itself progresses towards higher and higher orders of organization.

J. von Neumann explains that, paradoxically, the unreliability of the components of a living organism—as opposed to the reliability of those of a machine—is the condition of the organism's overall reliability.[35] The constant disorganization and reorganization, destruction, and reproduction of these components generates a high variety of possible responses to the disturbances a living being is bound to encounter in its internal or external environments. This is what H. von Foerster calls "order from noise."[36] I shall come back later to this fundamental aspect of the new cybernetics.

An open system exchanges energy and (possibly) information with its environment. It seems that the more complex a system is, the more important for its survival information inputs become, relative to energy inputs. Evolution is "a progressive substitution of information to energy."[37] In other words, only certain types of open systems exchange both energy and information, while all are open to flows of energy.

Thermodynamics is consequently not totally irrelevant to the study of social systems; on the one hand, it can serve to formalize actual energy exchanges taking place within these systems, and on the other hand it can suggest pertinent analogies. As far as the former aspect is concerned, economics, the science of energy transforming systems *par excellence*, has no place in its conceptual apparatus for thermodynamic variables. Of course, economists have always included in their models the circulation of raw materials and of fuels of various kinds. However, energy sources are accounted for only in monetary terms expressing opportunity costs. No consideration is given in any standard model of the production process to the thermodynamic efficiency of the latter. Such indicators are urgently needed; energy accounting is a new and promising field and one that is directly relevant to social indicators research, if one considers social indicators not as being complementary to, but including economic ones.[38]

As Georgescu-Roegen forcefully suggests, the *qualitative* nature of entropic transformation constitutes an inescapable conclusion—keeping in mind that the second law complements the first law which posits the invariance of the quantity of energy—despite attempts by Boltzman and others to reduce thermodynamic exchanges to probabilistic measurements. Now the formalization of qualitative changes calls for dialectical concepts, in the sense "that each concept and its opposite overlap over a contourless penumbra of varying breadth."[39] The sustained state of non-equilibrium of open systems, in other words the negentropy of life, and the entropy of inanimate matter are but the two sides of the same reality, namely organization. Information as a principle of organization (*in-formare* in Latin means to give form), presupposes the disorganization of the form to be.

Negentropy or Information

At the end of the 19th century, the notion of energy dominated science. Then the controversies associated with the problem of measurement at the subatomic level during the early years of this century confronted science with the elusive qualities of what we now call "information." But the formalization of this concept was not attempted before the late 1940s; C.E. Shannon's seminal work[40] established information theory on a solid, albeit narrow basis. In order to treat communication phenomena quantatively, Shannon left aside all considerations relative to meaning. For him, as for Wiener, information exists as such (I shall explain below why I wish to dissociate myself from this physicalist point of view); it is neither mass nor energy, even if both are necessary to its generation.[41]

Most definitions of information in a quantitative sense are consistent with the following one: the amount of information contained in a message is measured by how much it reduces the number of possible outcomes of the problem situation faced by the receiver of the message. This can be expressed by the logarithmic law: $I=K \text{ Ln } P$, where I is the amount of information contained in a message reducing the number of alternatives to one, K a constant, Ln the natural logarithm and P the number of possible outcomes; or if I is measured in bits, $I=\log_2 P$, a bit being the amount of information required to reduce the number of alternatives by half.[42]

Simply put, information theory deals with the conservation of information in a communication process. Not all the information sent is received—there is some "equivocation"—and not all the information received is what was sent, due to "noise" perturbation the transmission of a message. A typical problem, then, is the determination of the level of "redundancy," or surplus of information, capable of offsetting these disturbances at a minimum cost, and given the limited capacity of the channel being used. It is clear that here "[n]othing is said about whether the information is true, valuable, understood or believed."[43]

Valueless as it may be, information, in the absolute sense of the term as defined above, is a useful means of describing an open system. It was not long before several authors, Leon Brillouin first among them, were to point out the striking analogy between information and entropy (Boltzmann-Gibb's definition of entropy: $S=K \text{ Ln } P$).[44] Indeed, "information must be considered as a negative term in the entropy of a system; in short information is negentropy."[45] Conversely, the entropy of a system is a measure of the lack of information about it.

Perhaps one of the most steadfast obstacles to societal change is to be found in the insufficient development of social reporting. "Information is power" really means information is the power to transcend routinized social interactions. Shannon's "information" is not necessarily socially relevant, but its existence or nonexistence, the presence or absence of communication of any kind, constitutes a determinant factor in the viability of a living or social system. Anatol Rapoport suggests that information theory could be helpful in analyzing the information-gathering procedures of large organizations (e.g., government departments).[46]

Important as such considerations may be, "[they] open elegant formal perspectives, but [have] never led to any really operational results, neither in the social science nor in biology"[47] because of their extreme abstractness. In political science the clearest example of such an approach is K. Deutsch's work, about which L.C. Mayer comments:

> The crucial distinction between Deutsch's paradigm and that of Easton is that the concept of message units or informative bits in Deutsch's system replaces the less precise concepts of demands and supports in the Eastonian paradigm. This is crucial for the development of scientific knowledge . . . in that information can, in principle, be counted and measured.[48]

Let us emphasize "in principle," for Deutsch's call for the measurement of political communications has not been heard by many.[49] A further development of social reporting would, however, make the study of political communications more topical. At the very least social indicators are discrete bits of information, regardless of their meaning. It matters whether the unemployment rate is 4 percent or 14 percent, although these two indices carry an equivalent amount of information. But it also matters whether the reported rate is reliable or simply "noise," as it seems is the case at present in certain extraordinary circumstances (pockets of very high unemployment in regions like Newfoundland, Pennsylvania, Sicily, etc.).

In genetics it is undeniable that information theory has been a catalyst of sorts providing biologists with new conceptual tools enabling them to make a decisive first step toward a fuller understanding of the "genetic code." Progress beyond that point, nevertheless, has required a much more sophisticated conceptualization of the content as well as of the structure of genetic information.

One such refinement is Laborit's distinction between "circulating information" and "information-as-structure."[50] The former, exchanged between the various levels of organization of any open organic system, corresponds closely to Shannon's concept. The latter, however, concerns the overall laws that organize discrete elements (bits of information in the context of Shannon's theory) into organic wholes.[51] In the case of a living being, the genetic code correponds to the second definition. According to Laborit, the most complex systems are also the most open to circulating information but, conversely, their "information-as-structure" tends to appear rigid and the systems are closed ones from that point of view. In particular, Laborit laments man's incapacity to open himself on the societal level.

> The whole of man's plight comes from the fact that he has still not found the means to include [his own closed individual structure] in the larger ensemble whose finality would be his as well as that of all other men.[52]

Obviously, in living and social systems, the content of the message (*signifié*) as well as the message itself (*signifiant*) need to be taken into account. But I will show hereafter that, on the one hand, this most common critique of information theory fails to raise an even more fundamental question—one that E. Morin discusses in a quite novel way. On the other hand, to analyze the *signifié* it is necessary to distinguish between several more or less significant aspects of it.

Here we may remind ourselves that it takes two to communicate. Moreover, the original technological problems which led to the elaboration of information theory clearly involved a human source and/human receiver besides a machine. Communication is essential, or at any rate from the point of view of immediate experience, a social phenomenon. Rare are the instances in which the natural sciences rely so explicitly upon an analogy drawn from the anthropo-social realm. (Of course, the contrary is quite common.) Intrigued by this circumstance, E. Morin asks how we can understand the "transdisciplinary" journey of concepts such as communication, information, code, message, inhibition, etc., whose social connotation is so obvious.[53] The fact that Shannon's conception of information is blind to meaning does not strike Morin as being as serious as a less often noticed, but more important deficiency: "Shannon's information theory occults the anthropo-social system it implies and within which it acquires meaning."[54] In particular, information can only be degraded, but cannot be regenerated (in the context of this theory); this is so because information is taken as a given, and nothing is said about its generation, although this is the most crucial problem for natural and social scientists alike. Only by trying to answer that question can we avoid reifying information. The irony is that the concept of information, rendered more unequivocal, and hence more appealing, by its detour through technology and cybernetics, comes back now to the anthropo-social field as a master-concept. Yet, "it has lost the relational and contingent characteristics which made it worthwhile."[55]

The subjective quality of information has also been noted by J. de Rosnay who suggests that, while negentropy is an objective reality, information is the product of the translation of that reality into symbolic terms by a human observer. And thinking subjects can use information in order to recreate negentropic processes. But, as F. Varela points out,[56] information is a dispensable means of describing living (i.e., self-reproducing, self-organizing) systems since, by definition, they do not depend on the observer's creative abilities to regenerate their rate of entropy production. When using concepts like message, code, input, output, etc., it is important to realize that *we* define the meaning of these terms in order to communicate specific instructions to the systems that we have designed. But living systems, including politically organized communities, cannot be properly mapped by such inadequate concepts; the latter remain nevertheless useful in situations where we wish to exercise control over such systems (e.g., the regulatory function of bureaucratic agencies).

Taking into account the content (*signifié*) of the message brings to higher levels of complexity. Nothing relevant can be said about the content of all types of social communications in a sweeping statement, but Jacques Attali's hierarchy of meaning suggests interesting directions of research,[57] and it is consistent with my own hierarchy of complexity.

In addition to the a-significant aspects of information which Attali treats like Shannon, he distinguishes between cybernetic information or signal, semantic information or discourse, semiologic information or symbol, and unconditional information or relation. An open system relies on signals to maintain a steady state; but the determination of the norms defining that steady state involves man's cultural and technical knowledge. In Jacques Attali's terms, language is a set of semantic information structuring society and man's relation to nature.[58] Only signals can be measured, yet these first two levels of meaning are proximate enough that they can be regarded as different aspects of what we define hereafter as cybernetic complexity, taking cybernetics in a rather broad sense.

The other two levels of meaning correspond very closely to what I hereafter call self-organizing complexity, especially the highest one, "unconditional information," the product of "man's capacity to create from the codes at his disposal complex, unexpected and non-codified information [and] to exchange it with other men (and other groups) unconditionally, without intervening definite codes.[59] That is to say, in particular, the capacity to relate to other people without the mediation of objects. Technologically advanced societies, however, have been able to incorporate into objects increasing amounts of information, to the point where objects have become an indispensable means of communication. What Attali describes as the transition to unconditionality constitutes a revealing illustration of self-organization, but it does not exhaust the subject.

Even though Attali does not explicitly mention J. Habermas in the context of his discussion of information, the similarities between their writings on communication is striking. According to T. McCarthy, Habermas' theory of "communicative competence" culminates in the notion of an ideal form of communication "free from all constraints of domination, whether their source be conscious strategic behavior or communication barriers secured in ideology or neurosis."[60] For Habermas, such conditions are necessary for the formulation of true statements. This he sees as the beginning of an answer to the question of the interdependence of truth and freedom. Viewed from this angle, Attali's "unconditional information" acquires a more considerable epistemological weight.

Concluding Observations

There is no obvious way of characterizing a complex system. The concept of system is a short-cut for evoking relations among interdependent and yet autonomous units whose operations are mutually compatible and result in the production of organized patterns of energy and information processing. These, it is important to underline, result in irreversible changes over time. As a system evolves, it becomes more dependent on its cognitive functions. There are formal ways of analyzing the information content of the messages exchanged between what an external observer defines as the system and its environment. However, in the case of human systems, partly because there is no objective way of separating the environment from the system, and thus no objective way of treating information without interpreting its meaning—which is another way of saying that *we* define what the structures of societal communication are—the qualitative and subjective aspects of information exchanges are the most important. In particular, from a political standpoint, the question of the control over communication processes by the various elites and power structures, and the possibility of engaging in public debates where all interests can be heard is, or ought to be central to contemporary democratic theory. I shall come back to many of these points in subsequent chapters.

A recent example of the implications of complexity and of the strategic importance of communications in socio-political systems can be found in the circumstances surrounding the accident at the nuclear power plant at Chernobyl, near Kiev in May 1986. Technological, bureaucratic, ideological, and diplomatic factors all played a part in this human drama. It seems rather obvious that, even in the case of the Soviet Union, state institutions are not monolithic entities comparable to super-individuals, but complex systems within which a variety of actors pursue a multitude of strategies the outcome of which is truly systemic though (i.e., a relative inability to cope with an unexpected situation requiring co-ordinated responses). Hypothetically, information theory could be helpful for analyzing the loss of information that seems to have occurred between the local and central authorities; but the true measure of the complexity of this event can be gauged only in the light of a critical reflection on the constraints imposed by the prevailing ideological norms on the communication of undesirable and threatening messages in a "closed society."

Notes

1. Szymon Chodak, *Societal Development* (New York: Oxford University Press, 1973), pp. 42–76.
2. Edgar Morin, "Complexity," *International Social Sciences Journal*, Vol. XXVI (1974), No. 4, pp. 555–556.
3. See N. Georgescu-Roegen, *The Entropy Law and the Economic Process* (Cambridge, Mass.: Harvard University Press, 1971), Chapter II; Stafford Beer also stresses the relationship existing between complexity and uncertainty, see S. Beer, *Decision and Control* (London: John Wiley, 1966), p. 258.
4. See Chapter 4.
5. *The Entropy Law and the Economic Process*, p. 45; note that here I quote directly from the author whereas the previous argument concerning the adequacy of a logic of classes is my own, even if inspired from the same passage.
6. Warren Weaver, "Science and Complexity," *American Scientist*, Vol. 36 (1948), pp. 536–544; Weaver's paper is conveniently summarized in Ronald Brunner and Garry Brewer, *Organized Complexity: Empirical Theories of Political Development*, (New York: Free Press, 1971), pp. 102–103, see also T.D. LaPorte, "Organized Social Complexity: Explication of a Concept," in T.D. LaPorte (ed.), *Organized Social Complexity* (Princeton, N.J.: Princeton University Press, 1975), p. 5. My account of Weaver's arguments is drawn from these sources.
7. W. Weaver, "Science and Complexity," p. 539, quoted by Brunner and Brewer, *Organized Complexity*, p. 103, the emphasis is Weaver's.
8. Using Erich Jantsch's vocabulary, see E. Jantsch, *Design for Evolution* (New York: Braziller, 1975), Chapters 2 and 3 in passim.
9. *Op. cit.*, pp. 556 and 557.
10. See Chapter 8.
11. "Complexity," pp. 558 and 559.
12. According to Hall and Fagen, "A system is a set of objects together with relationships between the objects and between their attributes" (p. 81); they continue by stating that ". . . the term 'system' is used only when some degree of wholeness exists" (p. 85), and they call "degenerate systems" sets of parts with complete independence; see A.D. Hall and R.E. Fagen, "Definition of System," in Walter Buckley (ed.), *Modern Systems Research for the Behavioral Scientists* (Chicago: Aldine, 1968), pp. 81–92.
13. A. Angyal, "A Logic of Systems," in F.E. Emery (ed.), *Systems Thinking* (Penguin Books, 1969), pp. 17–29.
14. L. von Bertalanffy, "The History and Status of General Systems Theory," in G.J. Klir (ed.), *Trends in General Systems Theory* (New York: John Wiley, 1972), pp. 21–41; this, of course, does not imply a return to Aristotle's conception of science as such, nor even the necessary acceptance of any part of it.
15. Hans Jonas, *The Phenomenon of Life* (New York: Harper and Row, 1966), p. 201.
16. Lancelot Law Whyte, "Introduction," in L.L. Whyte (ed.), *Aspects of Form* (New York: American Elsevier, 1968), p. 2.

17. See ibid., in passim; F. Jacob remarks that "All the coordination of the cell is . . . dependent on the geometrical deformation of a few proteins . . . ," F. Jacob, *The Logic of Life* (New York: Pantheon Books, 1974), p. 301.

18. René Thom, *Structural Stability and Morphogenesis* (Reading, Mass.: Benjamin, 1975).

19. R. Thom discusses the relevance of his "catastrophe theory" to these first two problems, see ibid., pp. 280–324.

20. Which is not to say that the variety of possible qualitative transformations is infinite; these may take place in accordance with some general principle of morphogenesis, but the latter is not synonymous with infrastructural regularity implying continuity.

21. Jacques Sauvan, "Tombeau pour Antée: l'événement et le schème ou l'exigence de sécurisation," *Communications*, Vol. 18 (1972), pp. 122–127. Of course, the Second Law of Thermodynamics contradicts this quasi-cyclical cosmology; the latter, however, has been downplayed, ignored, or even explicitly rejected by many scientists for a long time, and, in many cases, is of little or no practical consequence for researchers working in extremely narrowly defined subfields.

22. Edgar Morin, "Le retour de l'événement," ibid., p. 8 (my translation).

23. Bertram M. Gross, "Systems Mapping for the Emerging Nation City," in M.D. Rubin (ed.), *Systems in Society* (Washington D.C.: Society for General Systems Research, 1973), p. 205.

24. K. Popper, *Objective Knowledge* (London: Oxford University Press, 1972), p. 292.

25. Ibid., p. 294.

26. J. Agassi, "Methodological Individualism," in John O'Neil (ed.), *Modes of Individualism and Collectivism* (New York: St. Martin's, 1972), p. 187.

27. P. McCormick, "The Concept of the Self in Political Thought," *Canadian Journal of Political Science*, 12:4 (1979), pp. 689–725.

28. See Roberto M. Unger, *Knowledge and Politics* (New York: The Free Press, 1975), Chapter 1.

29. A. Etzioni, *The Active Society: A Theory of Societal and Political Processes* (New York: The Free Press, 1968), p. 2.

30. E. Morin, "Complexity," *International Social Science Journal* (1974), No. 4, p. 558.

31. D. Easton, *A Systems Analysis of Political Life* (New York: John Wiley, 1965), pp. 17–18.

32. See I. Prigogine and I. Stengers, *Order Out of Chaos: Man's New Dialogue With Nature* (New York: Bantam Books, 1984), p. 104.

33. N. Wiener, *The Human Use of Human Beings: Cybernetics and Society* (2nd edit., Garden City, New York: Anchor Books, 1956), p. 28.

34. The following works, although quite different with regard to their purpose and content, represent two valuable attempts to deal with the problem of entropy in the social sphere: N. Georgescu-Roegen, *The Entropy Law and the Economic Process* (Cambridge, Mass.: Harvard University Press, 1972), and Stephen Coleman, *Measurement and Analysis of Political Systems* (New York: John Wiley, 1975).

35. J. von Neumann, *Theory of Self-Reproducing Automata* (Urbana, Ill.: University of Illinois Press, 1966); see also E. Morin, "Complexity," pp. 560–561.

36. H. von Foerster, "On Self-Organizing systems and their Environment," in M.C. Yovits and S. Cameron (eds.), *Self-Organizing Systems* (New York: Pergamon, 1960), pp. 31–50. See Chapter 3 for an elaboration of this point.

37. Yves Barel, *La reproduction sociale* (Paris: Anthropos, 1973), p. 95 (my translation).

38. For examples of such research, see M. Slesser, "Energy Analysis in Policy-Making," *New Scientist* (1 Nov. 1973), pp. 328–330; R.S. Berry and H. Makino, "Energy Thrift in Packaging and Marketing," *Technology Review* 76 (Feb. 1976), pp. 32–43; E. Hirst, "Food-related Energy Requirements," *Science* 184 (12 Apr. 1974), pp. 134–138; D. Pimentel, "Food Production and the Energy Crisis," *Science* 182 (2 Nov. 1973), pp. 443–449; for a summary of these works, see J. de Rosnay, *The Macroscope* (New York: Harper and Row, 1979), Chapter 3. Briefly stated, it appears that economic efficiency, that is to say high productivity, while beneficial in monetary terms, is quite costly in energy terms.

39. N. Georgescu-Roegen, *The Entropy Law*, p. 14.

40. C.E. Shannon and W. Weaver, "A Mathematical Theory of Communication," *Bell System Technical Journal* (1948); *The Mathematical Theory of Communication* (Urbana, Ill.: University of Illinois Press, 1949).

41. Which is to say that information has a physical existence, but one that cannot be reduced to the concepts of classical physics, see E. Morin, *La méthode* (Paris: Seuil, 1977), p. 307.

42. For more details, see G.A. Miller, "What is Information Measurement," in W. Buckley (ed.), *Modern Systems Research for the Behavioral Scientists* (Chicago: Aldine, 1968), pp. 123–128, and Leon Brillouin, *Science and Information Theory* (New York: Academic Press, 1962 second edition).

43. G.A. Miller, ibid., p. 124. However D.M. MacKay does not agree with this generally accepted view; according to him, the mechanical message does not recover its meaning in the mind of the receiver, but is itself meaningful insofar as it allows the recipient, without further reinterpretation, to select a course of action from the range of his "states of conditional readiness for goal-directed activity," see D.M. MacKay, *Information, Mechanics and Meaning* (Cambridge, Mass.: The MIT Press, 1969), pp. 19–30.

44. Leon Brillouin, *Science*, p. 1.

45. Leon Brillouin, ibid., p. xii. A. Koestler speaks ironically of the negative turns of phrase used by information theorists in order to avoid "falling into the heresy of vitalism, of reverting to Aristotle's entelechies, Leinbniz's monads or Bergson's *élan vital*," in *The Ghost in the Machine* (Chicago: Gateway, 1970), p. 200.

46. See Anatol Rapoport, *General Systems Theory: Essential Concepts and Applications* (Tunbridge Wells, Kent: Abacus Press, 1986), p. 161.

47. Jacques Attali, *La parole et l'outil* (Paris: PUF, 1975), p. 68 (my translation).

48. Lawrence C. Mayer, *Comparative Political Inquiry* (Homewood, Ill.: The Dorsey Press, 1972), p. 136.

49. See Karl Deutsch, *The Nerves of Government* (New York: The Free Press, 1966, second edition); there is, however, at least one noteworthy attempt to comply with Deutsch's exhortations, namely Richard W. Chadwick, "Steps Toward a Probabilistic Systems Theory of Political Behavior, with Special Reference to In-

tegration Theory," in H.R. Alker et al. (eds.), *Mathematical Approaches to Politics* (San Francisco: Jossey Bass, 1973), pp. 67–112. Also, for applications of the concept of entropy to the measurement of political relations, see Stephen Coleman, *Measurement and Analysis of Political Systems*; and L. Cohen and P. Warwick, *Political Cohesion in a Fragile Mosaic: The Yugoslav Experience* (Boulder: Westview Press, 1983).

50. See Henri Laborit, *La nouvelle grille* (Paris: Laffont, 1974), pp. 18–40.

51. Such a spatial dichotomy, so to speak, could also be interpreted in a temporal dimension. Instantaneous exchanges between a source and a receiver corresponds to "circulating information" and invariant, perhaps even internalized, potential forms are equivalent to "information-as-structure."

52. Ibid., p. 38 (my translation).

53. See the introduction by I. Nahoun to Edgar Morin, *L'Esprit du temps*, vol. 2 (Paris: Grasset, 1975), p. 19.

54. E. Morin, *La méthode*, p. 304.

55. Ibid., p. 311.

56. F. Varela, *Principles of Biological Autonomy* (New York: Elsevier, 1979), Preface and Chapter 9 in passim.

57. Jacques Attali, *La parole*, pp. 62–85.

58. Ibid., pp. 73–76.

59. Ibid., p. 78 (my translation).

60. T. McCarthy's introduction to J. Habermas, *Legitimation Crisis* (Boston: Beacon Press, 1973), p. xvii; McCarthy subsumes here arguments developed previously by Habermas; these have now been articulated into a vast conceptual framework in his *The Theory of Communicative Action*, Volume 1: *Reason and the Rationalization of Society* (Boston: Beacon Press, 1984).

2

Cybernetic Complexity

Cybernetic complexity holds a pivotal position in the threefold hierarchy of complexity unfolding through these first three chapters, for it reintroduces the notion of closure as an operational category. Obviously, the systemness of a system implies a totality akin to self-closure but, as we have seen, the perpetuation of this unity is made possible only on the condition that the system evolves more and more effective ways of interacting with its environment. Enduring systems are also open systems. The regulation of the flow of inputs and outputs implies, in return, a certain kind of closure. Controlled systems of the type that are described in cybernetic models are indeed open to flows of energy but comparatively closed to information. Absolute closure is impossible to achieve, since the Second Law of thermodynamics rules out perpetual motion, and, after all, even messages must have an energy support. What this means, however, is that information about the system states and/or the position of the system vis-a-vis some part of its environment circulates within a closed loop; again, it does not at all denote a total isolation, as would be the case if the system did not receive any input of information, let alone of energy.[1]

No matter how crucial communication processes are to an understanding of the behavior of complex systems, the theory of complex systems cannot be limited to the study of information flows. A serious weakness of information theory is that it does not differentiate between the signals transmitted by information channels. However, the unfocused study of communication processes needs to be complemented by an inquiry into the effects of information-as-command. This is precisely the object of cybernetics, a scientific discipline which was originally conceived as the "Science of control in the animal and the machine" (Wiener) and since then has become a fad: "feedback" has come to replace "reaction" in the English language.[2] Thus it is appropriate to begin this chapter by taking a critical look at the basic concepts of cybernetics.

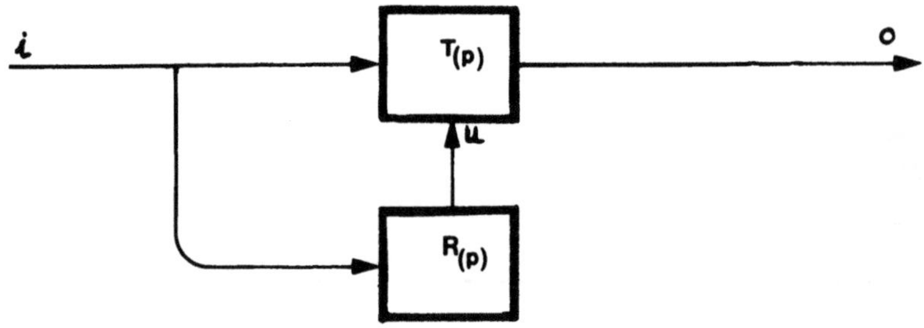

Figure 2.1. The feedforward loop.

I argue that the political relevance of cybernetic theory does not lie in the manipulative applications of control theory but rather in the epistemological significance of the kind of closure which is characteristic of system structures including at least one feedback loop. Their *self*-regulatory capabilities suggest that they possess an emerging form of autonomy, that they constitute at least embryonic "beings." Yet the first generation of cybernetic models hardly took this dimension into account—as opposed to those of the new cybernetics which are discussed in the next chapter.

Of Feedback and Variety

There are two ways to build regulatory processes into a system. One is to design a "feedforward loop" by means of which information about the expected discrepancy vis-a-vis the desired goal state is fed into the system in advance (see Fig. 2.1). Feedforward regulation can be observed in a large number of systems (e.g., preventive medicine in the control of public health). This process implies a complete knowledge of the behavior of the system that needs to be regulated, under all possible conditions. Such a prerequisite is almost impossible to meet in situations of extreme complexity.

More prevalent, at least in living and social systems, are a posteriori regulatory mechanisms which dampen or amplify fluctuations once they have occurred. Such are, indeed, the effects of a closed feedback loop (see Fig. 2.2).[3]

Let $F(p)=o/i$;[4] if the feedback is negative, i.e., minimizes the differences between actual and desired or "normal" system states, we also have $F(p)= sT(p)/(s-sR(p)T(p))=T(p)/(1-R(p)T(p))$, or, if the feedback is positive, i.e., amplifies these differences, $F(p)=T(p)/(1+R(p)T(p))$. Note that, if $T(p)$ is of a magnitude largely superior to one, which is quite possible, then $F(p)$ tends toward $1/R(p)$. This shows clearly that the feedback network, rather than the forward network, dominates the controlled system. Now it is not necessary to know the functions $T(p)$ and $R(p)$ in order to analyze the

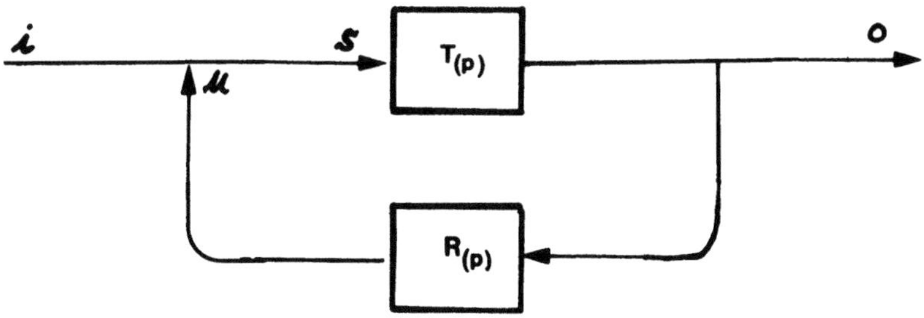

Figure 2.2. The feedback loop.

behavior of a controlled system. They may be too difficult to solve anyhow. In practice, the way to bypass analytical solutions is to resort to step-by-step simulations of the behavior of the system under consideration. Ashby points out that biologists and economists, among others, must often use mathematically unsolvable equations.[5] To overcome this difficulty, he writes all his transformation functions in a tabular form, setting aside the question of their analysis.

For Ashby the interaction between two machines or classes of systems provides the most general model for the description of feedback processes. Let us consider the interaction between the quality of life and the determination of political priorities symbolized respectively by Tables 2.1 and 2.2, which show the transformation functions characteristic of those systems. We will suppose that we observe the relationships between the states of one system and the parameters of the other which are laid out in Table 2.3. We have then this equation, starting from the vector (VP,C).

$$(VP, C) \rightarrow (P, NC) \rightarrow (P, C) \rightarrow (A, NC) \rightarrow (A,C) \rightleftarrows (G,C)$$

Quality of Life

	from Very Poor (VP) to:	from Poor (P) to:	from Average (A) to:	from Good (G) to:
Policy #1	P	A	G	A
Policy #2	VP	P	A	A
Policy #3	VP	VP	P	A

Table 2.1

Political Priorities

	from Non-Conservationist (NC) to:	from Conservationist (C) to:
Attitude #1	C	C
Attitude #2	C	NC
Attitude #3	NC	C
Attitude #4	NC	NC

Table 2.2

In other words, the global system reaches some kind of equilibrium, such that the quality of life never deviates from the average and good positions.

We have not, so far, brought feedback to bear upon the problem of regulation, that is to say the capacity of a system to maintain a set of crucial variables within a range of "acceptable" values in the face of external disturbances.[6] In the case of the above example, other factors than political preferences, and upon which the policy has little or no control, determine the quality of life. These factors are called disturbances; Figure 2.3 describes the resulting pattern of interactions. Table 2.4, which now replaces T_2, determines the type of policy to implement for each political priority and each kind of disturbance. Starting with the vector (VP,C,I), and supposing

T_1	Quality of Life:	VP	P	A	G
	Type of attitude:	2	2	1	2
T_2	Priority:	NC	C		
	Policy No:	2	1		

Table 2.3

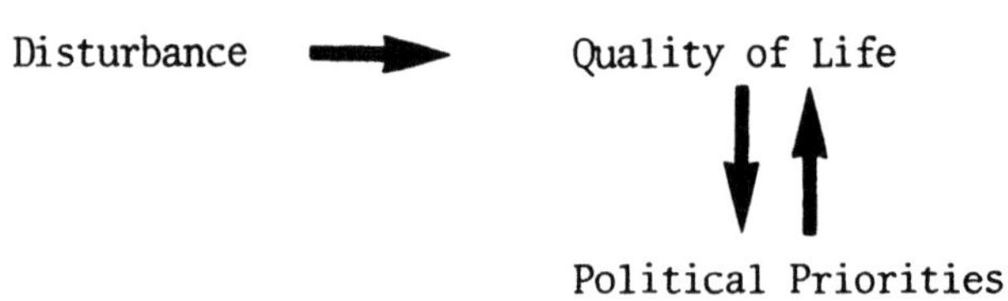

Figure 2.3

Political Priorities

	NC	C
Disturbance #I	3	2
Disturbance #II	2	I
Disturbance #III	3	I

Table 2.4

that disturbance I produces its effects over a relatively long period of time,[7] we observe:

$$(VP, C, I) \rightleftarrows (VP, NC, I)$$

Thus, in the event of disturbance I, the political system has no control over the quality of life, which remains very poor.[8] As this simple, if not simplistic, example clearly shows, the mere existence of a feedback loop does not guarantee perfect regulation. In search of an analysis of the condition of regulation, we must again turn to Ashby whose "law of requisite variety" is of paramount importance for the design of any viable system, and has serious, although rather neglected, implications for politics.

Let us first define variety: it is a measure of complexity, which is proportional to the number of parts as well as their relations in a given system.[9] Now, a regulator R regulates another system S if, and only if, it can react appropriately to the disturbance D to which S is subjected. The result is that R causes the variety of S's outputs to decrease. For instance, to maintain a national currency within a limited range of exchange values, a central bank must reduce the variety of the quotations on the money market. What Ashby has established is that R can limit the variety of S's outcomes by no other means than its own variety, or, as he puts it, "only variety can destroy variety."[10] Therefore, R will control S fully at the condition that R's variety matches that of D. This is the law of requisite variety. It is of universal import, pertaining to regulation as a general phenomenon, and regulation is obviously one of the dimensions of politics. (I come back to this discussion of politics as cybernetic control in Chapter 6.) Political science, unfortunately, has paid little attention to Ashby's law.

Stafford Beer, however, has written extensively about the organizational and political implications of Ashby's law.[11] For Beer, the internal and external variety of any social organization is such that management can never hope to match it. This problem is considerably more acute for the political authorities. Societal regulation always entails, therefore, some effort on the part of the authorities to reduce the variety generated by the systems with which they interact. They can refuse to see things as they are and "manage a world of their own," populated by meaningless statistical aggregates; they can even constrain individual freedoms by force.[12] Not only is this a well-known human tendency, but some degree of variety reduction is unavoidable if Ashby's law is not to be violated. As a matter of fact, an interesting research assignment would consist in developing a set of criteria for measuring variety reduction due to bureaucratic organizations. Does this mean that cybernetics provide a scientific justification for the robotization of man? The answer to that question is definitively negative, despite widespread misconceptions to the contrary.

To reduce the variety generated by the controlled system is only one of the possible means of satisfying Ashby's law; the other is to amplify the variety of the regulator. From the standpoint of protecting human rights, the second alternative appears preferrable to variety reduction; it is also advisable from a cybernetic point of view, insofar as a complex regulator is more likely to react adequately to unexpected disturbances. Admittedly, to amplify the variety of socio-political regulators may involve excessive costs. At the limit, it could mean assigning a guardian-angel-turned-bureaucrat to each and every citizen!

In a given society, the most acceptable balance between variety amplification and reduction can only be determined politically. There are, however, certain objective constraints which must be taken into account. S. Beer argues that technology has been mobilized "on the wrong side of the variety equation."[13] It has made the world more complex than need be. Computers have considerably increased the amount of information absorbed by modern organization whose variety, however, has remained constant. Used as they are, computers become alienating. According to Beer, we ought to employ them on the other side of the variety equation, where they could amplify the regulating variety of our organization, particularly of our public institutions. This involves the careful design of simulation models retracing the "pattern by which variety in the system is deployed."[14] Beer does not minimize the difficulties inherent in such a task, but he insists that his brief experience in Chile under President Allende proves that it can be done.[15] Instead of memorizing and cross-tabulating more and more irrelevant data, as they do at present, computers would, thanks to these simulation models, screen out incoming information and discard most of it since it would fit expected variations.

With the help of the remaining set of indicators pointing to potential breakdown in the system, decision-makers, at various "levels of recursion," should be able to intervene when necessary and to explore a broad range of policy alternatives.[16] They would dispose of a higher regulating variety. Note that Beer is describing a multi-level process; he is not arguing that political regulation should be entrusted to a centralized bureaucracy. This would be in direct contradiction with Ashby's law, in any case.

Taking the nervous system as an ideal type of adaptive organization and reasoning by analogy, Beer stresses that a viable system achieves a balance between centralizing and decentralizing tendencies. The command center cannot operate without the assistance of autonomous subsystems linked together in a horizontal network.[17] Similarly, information screening and corrective actions ought not to be carried out only at the top level of any organization or societal system but, on the contrary, at any level where the resources correspond to the need.

The principle of requisite variety and Beer's related concept of "relaxation time" help us to understand K. Deutsch's choice of load, lag, gain, and lead as the variables best describing the performance of controlled systems.[18] Obviously, a system is overloaded with information when its regulating variety is too low. (In many countries, the judicial system is chronically overloaded.) The lag in the response of a system to a threat corresponds to what Beer calls "relaxation time." It varies with both the internal variety of the controlled system and the extent to which incoming messages are contemporaneous with their sources. In a way, the delay characteristic of a message is a measure of its quality. Beer's insistence on the need for "real-time" communication networks in socio-political system may be a little naive, seeming to minimize the importance of nontechnological obstacles. Yet it should not be ignored.

Gain is just the other side of the coin: the higher the regulating variety and the lower the relaxation time of a system, the more appropriate its gain will be, i.e., the nearer to the mark each corrective step taken by the system. The concept of lead, however, raises a different question. The capability manifested by complex systems to anticipate disturbances, and thus to transcend the simple stimulus-response schemata, corresponds to the process of variety amplification discussed above.[19]

It is now time to discard once and for all the simple model used above to illustrate the concepts of feedback, limited as it was to the interactions between a single regulator and a single effector, and to consider systems including many such subsystems. R. Landry and V. Lemieux suggest that each and every regulator or effector is itself composed of a regulator and an effector interacting through a maze of feedback networks.[20] Provided that the connections between the subsystems are not too "rich"—in Ashby's terms—they can adapt to each other. What this means is that the parameters specifying internal connections are gradually selected by adaptive learning, instead of being fixed as was the case in our earlier model. When external disturbances—even unexpected ones—excite a given subsystem, the others will adapt to this change, and stability is thus restored. This is why Ashby calls such systems "ultrastable" machines. Homeostasis is the process through which ultrastability is maintained. Automated plants offer an example of homeostatic systems; some public agencies performing specific and standardized tasks (e.g., Post Office) also conform to this model.

However, it is important to underline that the usefulness of the concept of ultrastability depends on the existence of only weak interactions within the system as a whole. According to Ashby, "Examples of environments that are both large and richly connected are not common, for our terrestrial environment is widely characterized by being highly subdivided."[21] In other words, most complex systems are "nearly decomposable" ones, close to being made of independent systems, and not so complex after all. H. Simon, among other authors, pretends that nearly decomposable systems can be studied as totally decomposable ones, for short periods of time. It does not take long, however, before this approximation becomes the source of significant errors, even in the case of poorly interconnected subsystems, as D. Metlay has shown.[22] Of course, Ashby does not go so far as to reduce large systems to an aggregate of almost independent entities. But does his assumption regarding the weakness of internal connections in natural systems hold true for most man-made systems? LaPorte, Metlay, and some other writers answer negatively; they stress the richness of the internal connections characteristic of societal systems.[23] But, from a normative point of view, Ashby's ultrastability as well as Beer's views on requisite variety imply that centralized methods of socio-political control are bound to become sources of uncontrollable instability because such methods unnecessarily create additional connections where there already are so many. Regulation for regulation's sake can only produce chaos. This is not an endorsement of the neo-conservative rhetorical attacks against "excessive" government regulation. But neo-conservative pundits have identified an interesting problem: When do discrete regulatory schemes, each one of which may be desirable in its own right, result in dysfunctional systemic rigidities?

Learning holds a central place in Ashby's work and in cybernetic theory generally. However, learning is far from being an unambiguous term. For the behaviorists, changes in the immediate or even remote (in space and time) environment of an organism trigger a response entirely determined by those changes. This response, in turn, "conditions," more or less directly, the future behavior of the organism, which has thus learned a new task or a new habit.[24] But, according to O.K. Moore and D.J. Lewis, the behaviorist model rests on an implicit acceptance of the teleological frame of reference, even if it explicitly excludes the notion of purpose.[25] A simple learning situation, like that of the rat-in-the-maze, is better represented by a feedback circuit, in which a control centre manifestly acts as a memory and goal-setter.

All learning processes cannot be subsumed under that simplistic representation. Deutsch's insightful discussion of the process of political learning, for instance, raises questions that go beyond the analytical power of even the concept of ultrastability; to wit, "the measurement [of] the speed at which an organization learns to learn" presupposes the definition of a hierarchy of norms transcending homeostatic stability.[26] The next level on that hierarchy is another kind of steering process that C.H. Waddington calls "homeorhesis."[27] It is observable in complex systems going through dynamic transformations, for instance, the embryo of any species. Such "progressive systems" follow developmental pathways, or "chreods," which can be visualized as valleys in a tridimensional space, but it should be kept in mind that actual systems usually function in multidimensional spaces. All the feedback networks within a given system shape the valleys down which its subsystems flow, and the system as a whole can be represented as an "epigenetic landscape." This is not simply an analogy, but a vivid way of expressing the topological concept of attractor surface. A valley tends to guide the trajectory of a concrete evolving system, but in no way determines its course; at some points, two valleys may branch off from each other and there is always a certain indeterminacy as to which course the system will take. In other places a valley may be very shallow, so that the system may wander away from it, and possibly fall into another valley. Through homeorhesis a system is led back to the valley where it belongs, although not at the same point where it left it, but further along. In other words, while homeostatic regulation tends to maintain certain crucial variables constant, homeorhesis ensures the stability of a formative process. Clearly, both modes of regulation embody dynamic principles, but the former involves only synchronological changes, the latter, by contrast, is essentially historical. To explain the relationship between homeorhesis and homeostasis, J. Piaget states that "The various chreods characterizing epigenetic development and their spatio-temporal regulation through homeorhesis end in a more or less balanced mature state characterized by its homeostasis."[28]

We cannot, however, state that such a sequence holds in all contexts, or even that there are no other ways for both modes of regulation to combine. The problem of the functional relationship between hierarchized processes remains open. Clearly, homeorhesis is a process observable not only in the "strategy of the genes," but in other contexts: Piaget points to the homeorhetic character of a child's mental development, and Waddington himself is aware of the relevance of his conceptual innovation for the study of societal regulation. For instance, the "life cycle" of a politician, or of a regime, follows its course through a series of contingencies, or chreods, shaped by the constantly changing popular support of various groups, without ever being stabilized at an optimum level.

The concept of homeorhesis present more than a simply descriptive value, it also suggests original direction of research. Two of them were identified by Waddington. First, there is the problem of knowing what goes on at the points where chreods split up into two or more other chreods, in Waddington's words: "What can one say about the question of whether the system goes down one branch, or the other, or breaks up and part goes down each?"[29] This apparently simple question raises a series of highly significant issues. We are here confronted with the discontinuity of dynamic phenomena, the discreteness and qualitative character of the natural and social orders, that science had ignored or down played until not so long ago. Experienced administrators and elected officials know that policy choices are very much like informed guesses; most of the time, it is impossible to forecast the influences that will bear on the implementation of a program or strategy. In attempting to formulate theoretical interpretations of these realities, several authors put a new emphasis on positive (i.e., deviation-amplifying) feedback, a concept which had been given very little place in cybernetic models of the first generation. Moreover, this concern for discontinuous phenomena opens up new avenues of cooperation between mathematics and the natural (or social) sciences, as in the case of René Thom's "catastrophe theory."[30]

Second, as Waddington puts it, "when we are confronted with an unknown system, how do we find out what the shape of the landscape is."[31] It is probable that the strategy for exploring attracting surfaces developed by mathematicians in this regard goes a long way toward solving the problem of learning how to learn raised by Deutsch and many other writers.

All these notions and approaches point to a *problématique* altogether different from that which originally characterized cybernetics, especially with respect to the time scale they involve. Rather than redefining cybernetics or employing expressions like "the second cybernetics" (M. Maruyama), I have decided to regard them as partaking of an entirely different order of complexity, and to study them separately in the next chapter under the heading of "self-organizing complexity." Before discussing these aspects, however, it seems appropriate to take a second look at cybernetic complexity proper, from a more detached and critical point of view. Much of what precedes was intentionally limited to fundamental cybernetics—my contention being that socio-political cybernetics must be grounded firmly on such a basis, otherwise it would amount to no more than pretentious pseudo-science. However, it is time now for refocusing our attention on more obviously political problems.

Cybernetics Revisited

E. Morin insists that closed and open states should not be considered as mutually exclusive but rather as complementary, even indissociable concepts.[32] A fully closed loop would result in a "vicious circle," and yet the type of causality associated with feedback processes is anything but vicious. The reason being, as F.G. Varela remarks, that "vicious" circularity occurs in a "duration-less time," whereas feedback processes develop through time; hence, to build a logic of closedness it is necessary to "introduce a new time-like dimension which defines a new domain."[33] Systemness corresponds, for Morin, to the notion of being, or, at any rate, to its most archaic form, namely the possession of an organizational capacity. In cybernetic theory, a machine is a being because its activities articulate its parts into a whole. It also enjoys a primitive form of perduring existence. Although belonging to the universal scheme of things and intimately related to its environment, they have achieved a certain degree of autonomy. This is the paradoxical result of openness manifested by the negentropy of life. Eventually, the homeostatic machine acquires an embryonic self, insofar as the closed feedback loops transform production activities common to all machines into a production of the producer, that is to say the production of a self:

> The productive recurrence of the same on the same, (re)producing itself and reproducing itself by itself, causes the emergence of a reality of an entirely different order expressed by . . . the concept of self.[34]

Contrary to what happens in living organisms, self-production is only "phenomenal" in the case of cybernetic artifacts. It consists in the reproduction of the retroactive totality but not of the being of the system and that of its subsystems.[35] For instance, a thermostat plays no part in the reproduction of the furnace, yet it is essential to the reproduction of the regulatory activity built into the system by its human maker. The point is that this essential difference should not hide from us the logical linkage existing between production and reproduction at all levels, even at the physical level. The concept of closure and thus of *self* provides the essential link. To the extent that actual or hypothetical machines and/or arbitrarily defined systems perform acts of production, they are also potentially capable of self-reproduction.[36] Morin's insightful analysis, and the convergent reflections of authors like G. Guenther, or F. Varela, take us away the mechanistic definitions of cybernetics which were common a generation ago. This should not come as a revelation today when Artificial Intelligence has forced a renewal of interest in the concept of "mind."[37]

We are now in a better position to comprehend the origin of the self of a social collectivity. In making the self of a social collectivity the carrier of an "active orientation," Etzioni comes close to the position developed above. He fails, however, to pursue it far enough considering that he does not explicitly address the question of a collective action feeding back upon itself as a logical problem in its own right.[38] According to Morin, the emergence of a self opens the door to freedom. A cybernetic regulator—which our author calls "*appareil*"—allows the whole being of the system to emancipate itself from the constraints of its environment, but at the cost of its subjection to the regulator's commands.[39] Cybernetic complexity reveals here its Janus-like twin faces. It is equally misleading to extoll the liberating potentialities of cybernetics or to condemn it for its coercive implications, if both perspectives are not kept in balance. I will try hereafter to preserve this dialectic tension, the political connotations of which should be obvious.

Beginning with the more attractive face of cybernetics, it is clear that feedback processes tend to isolate a system from the chaos of the external world. No longer at the mercy of all the disturbances that affect it, a controlled system can persevere in the pursuit of its own goals. The emancipating character of a cybernetic approach did not escape Karl Deutsch. In particular, Deutsch stresses the high degree of autonomy that self-steering organizations can potentially reach.[40] Our author goes on to suggest a parallel between Sartre's idea that "we act out our personality at every moment of our life" and what he perceives as the necessity for "every self-organizing system [to] remake its own memories and inner structure as it acts."[41] Hence, for Deutsch, self-steering systems are doubly autonomous: first, the system as a whole operates (relatively) freely in its environment, and second, the regulator is itself capable of manipulating its goals. In a similar vein, R.W. Chadwick defines politics as a "collective learning process which manipulates goals and actual conditions either against or through [expectations and disturbances]," and power as "the ability to attain goal-states against [these same obstacles]."[42]

For D. Easton, the most significant aspect of feedback processes lies in the fact that they establish a linkage between outputs and the sources of "specific support."[43] Starting from this comparatively narrow basis, Easton reaches a similar conclusion, namely that "feedback information" is "an essential condition for purposive innovations."[44]

The purposive nature of political life is thus quite explicitly acknowledged by these authors. As K. Deutsch puts it, "cybernetics offers not only a gain in technical competence but also a possibility of restoring to problems of purpose their full share of our attention."[45] However, what they say about the goals being pursued remains vague, and has given rise to the objection that the cybernetic approach is *status quo* oriented. The one-sided emphasis placed by Deutsch on the "self-preservation of autonomous organizations,"[46] lends credence to this argument. However, I feel compelled at this point to rescue cybernetics proper from what P. Birnbaum sees as a close association with the "end of ideology" movement:

> [for Deutsch, traditional ideologies] lose their indispensable character, since the political system now has become identified with a self-regulating mechanism, whose guidance does not require any "vision" (Wolin), nor any . . . moral or metaphysical world-view.[47]

Admittedly, Deutsch and Easton tend to downplay the importance of cultural and ethical values, insofar as these are internal factors which do not fit easily into their models of communication processes between the political system and its environment. Cybernetics has not been originally conceived to handle such norms, but rather to deal with more "objective" constraints, which prompts us to reassess the place held in it by the notion of purpose.

First generation cyberneticians surrounded themselves with many lines of defense when they suggested that teleological reasoning could become truly scientific under certain conditions. These conditions included the explicit rejection of any "final cause," and the strict adherence to a behavioristic approach focused upon input-output relationships, without consideration for intrinsic organizational factors. Thus "[t]eleological behavior . . . becomes synonymous with behavior controlled by negative feedback."[48] Such a claim was strongly attacked by several authors, including Richard Taylor, and Hans Jonas, on the grounds that only human beings possess a sense of purpose.[49] Also for W.P. Brandon, no semantic subterfuge should be allowed to hide this critical point.[50] There is here a confusion between the origin of the goals built into cybernetic machines and the reality of goal-seeking processes. In the same way that information theory disregards the origin of messages to deal only with their transmission, cybernetic theory—at least the "first cybernetics"—pays no attention to the question of knowing how the goals that controlled systems are designed to achieve were generated in the first place. Ashby has addressed himself to this question, but unfortunately reduces it to a discussion of the factors affecting the transmission of information between a "goal sponsor," a designer, and a machine.[51] That goals generally reflect a wide variety of values and constraints rather than being spontaneously improvised by any one actor in particular should not be overlooked, especially when cybernetics is applied to politics.

My contention is that to use such an epistemological myopia as a pretext for denying the reality of self-closure in complex systems would make any discussion of purposefulness hopelessly sterile. Indeed, the unique value of cybernetics is that it does not presuppose an arbitrarily defined threshold of purposefulness, but takes that dimension into account at the level where it emerges phenomenologically. Thereby, cybernetics establishes a bridge between the objective and subjective realms.

Coming back now to Birnbaum's charge, our author does not seem to realize that the development of cultural and ethical norms in a society throughout time depends upon an original self-closure originating in nature, which a cybernetic approach can help us to comprehend. Of course, the concept of feedback—and its corollary, that of autonomy—do not suffice to explore the normative space, but they are a key to it. To dispense with cybernetic concepts means renouncing to search for ways of escaping the positivistic separation of object and subject. Without a logic of closedness, the social science will continue to oscillate between social determinism and idealistic voluntarism; two philosophical positions which fail to establish a much needed "moral or metaphysical world-view" on a firm ground.

The former position precludes the consideration of self-referential variables, deemed to be "vicious" in the social realm, just as in nature. (It is, however, possible to profess a determinism limited to the subject-matter of the natural sciences.) Consequently, social determinism does not merely downplay the normative and creative aspects of politics, but altogether ignores them.[52] Determinism-cum-probabilism has more than tinged the behavioralist movement. As proselytizers of this movement, Deutsch and Easton could not have looked at G.S.T. and cybernetics as the building blocks of a new scientific paradigm, rather they envisaged them simply as means of making political science more "scientific." In that respect, Birnbaum's criticism is fully justified.

Like determinism, idealistic voluntarism denies the existence of natural phenomena exhibiting any degree of autonomy, but endows man with the gift of consciousness, thus pitting him against nature. Yet man's reflective abilities, his ethical values, and their potential concretization into collective projects, find in idealism only the slimmest foundation. Predicated upon the unexplained intervention in the individual or social psyche of an unknowable spirit, this philosophical perspective can too easily be turned upside down.

In brief, cybernetics, by itself, does not yield an imaginative "vision" transcending history, to use S. Wolin's terms.[53] However, by laying bare the logical and even physical prerequisites to the knowledge and practice of self-production, cybernetics helps us to comprehend freedom as a concrete reality. From there, and only there, imagination can relay science. But whose freedom? At this point, we must turn to the less attractive face of cybernetics.

A controlled system escapes a wide range of environmental constraints, and yet does not become, as a whole, fully autonomous. Emancipation through homeostasis also results in a subtler and more inextricable subjection to the programme of the regulator. A servo-mechanism controls a machine without ever being affected by the latter. In other words, the parts of a controlled system other than the regulator are not capable of autonomous reaction to the commands issued by the regulator. For instance, if the temperature has been set by mistake at 100°F on the thermostat dial, the vibrations of the overworked furnace, to say nothing of the cracking of the furniture or the gaspings of the cat, will not lower the selected temperature. According to Bateson, the operation of the regulator is determined by its own states as well as by the other parts of the system because the former must continuously adapt to difference in system states occurring in the latter. I do not challenge this argument, nor his conclusion that consequently the "mind" of the system is not located in the regulator, but is inherent to the system as a whole.[54] What I want to emphasize, however, is that in cybernetics communication processes are reduced to the mere processing of information-as-command. Thus applied to politics, cybernetics reintroduces the notions of will and purpose, but can hardly deal with ill-will, let alone rebellion or passive resistance. As we have seen, cybernetics has considerably enriched our conceptualization of the physical, biological, and social aspects of complexity. Because we can hardly do without it, we must also be keenly aware of its limitations and built-in biases.

In the aforementioned classification proposed by J. Attali, "cybernetic information" refers only to the transmission of signals (e.g., prices in economic exchanges); however, his next level, namely semantic discourse is just as relevant to cybernetics, and casts a new light on its bias. A program is always written in a language or code superseding all interactions within the system. It is accepted as a given.[55] In non-technological applications this perspective is dangerously misleading, yet it would be wrong to qualify it as necessarily "conservative." After all, it also suits a "radical" and messianic vision. Marxist rhetorics often ignore those sources of internal resistance which, blinded as they are by their "false consciousness," oppose the irresistible course of history (whereas Deutsch ignores the "progressive" forces digging the grave of a ruling class bent on maintaining the status quo).

In his attack on cybernetics, Morin fails to distinguish between computer programs, or algorithms, and what S. Beer calls "heuristics" which "prescribe general rules for reaching general goals . . . and do not typically prescribe an exact route to a located goal."[56] It would seem that heuristics goes some way towards the solution of cybernetics' dilemma. But let us now turn to another kind of criticism.

First, the question arises of the legitimacy of the application of cybernetics to politics. This objection can be set aside, for cybernetics was conceived originally as the science of control per se, and, by definition, it straddles traditional disciplinary barriers. But then, is not cybernetics itself guilty of the worst confusionism? Is it justifiable to hypothesize a structural identity between electrical circuits, living organisms, and social organizations? R. Tostevin sees precisely in the lack of "phenomenal regard" characteristic of the systems sciences' conceptualization of control the mark of their naive and potentially sinister belief in a natural philosophy that realizes "a will to control" as an end in itself.[57] This is a serious charge that F. Varela also levels against what he calls, using Rosenberg's phrase, the "gestalt of the computer."[58] He opposes it to his own understanding of information processes as being "constructive" rather than merely "instructive." But Varela does not consider it necessary to reject the computer gestalt altogether for it has its limited usefulness. What is required is the transition to a more reflective kind of cybernetic theory that fully recognizes the autonomous "will to control" their own domain of existence exhibited by complex (living) systems.

Cybernetics has been criticized also for its inability to conceive regulation as something else than a reaction to "errors."[59] Individually or collectively, we cannot always wait until something goes wrong to take a corrective decision. Yet, this is what the logic of the feedback process seems, at first glance, to imply when applied to a social context. Were it really the case, cybernetics would be little more than a gimmick contrived by technocratic sorcerer-apprentices to excuse away their bunglings! There are at least two reasons why this criticism is not as damaging as it appears to be.

To begin with, theory and practice do not necessarily coincide, and should not be confused, even though they can be closely linked.[60] For instance, the analysis of a policy problem by means of a computer simulation does not entail the implementation of the various alternatives considered. Thanks to computer simulation, all kinds of "deviations from the norm" can be visualized and tested without incurring any actual cost. In general, these simulation methods are inspired by cybernetic theory, and, to that extent, cybernetics helps to prevent, rather than simply justify, catastrophic policy decisions.

Let us probe deeper into the matter of "errors." No matter how unpleasant, or even horrifying, their consequences, human failings and physical breakdowns are parts of our experience. Thus it is crucial to distinguish between ethical norms and formal norms of stability. Cybernetics does not justify errors (a misplaced moral judgement), but acknowledges the tension existing between order and disorder in all realms, from the *physis*, where it was revealed by modern physics, to the polis, where it has been manifest since the end of absolutism. There is no unequivocal correspondence between order, or certainty, and the good, not even in Hobbes; nor is there an opposite one between disorder, or uncertainty, and the good, not even in Sorel. So to tag on cybernetics a label of moral irresponsibility because it shows how external disorder can be used to maintain internal order is a criticism that is more revealing of the critics' cosmology than being a valid philosophical refutation. Indeed, to attack cybernetics for its alleged apology of error denotes an unavowed longing for the beautiful, but illusive, simplicity of the clockwork cosmos or the harmony of an idealized polis that never was.

It remains that not all cyberneticians have been equally aware of the importance played by disorder in their models. Fewer go as far as to acknowledge the active presence of disorder within a controlled system (as opposed to its prevalence outside the system). I denounced earlier this blindness as being the sinister face of cybernetics. As E. Jantsch suggests, "organization *qua* stabilization may turn out to be disastrous in certain circumstances."[61] Examples of this abound. In the cultural field, for instance, attempts to resist the advance of a new scientific "paradigm" (in the Kuhnian sense) by trying to fit new ideas into prevailing models, where they do not belong, can only work against the development of scientific knowledge. Similarly, the last spasms of an unviable political regime (e.g., the slow retreat of the colonial powers from their overseas empires) are both futile and disastrous. To deal more completely with the paradox of order as a result of disorder, I now turn to "self-organizing" complexity.

Notes

1. In J.W. Forrester's words, "A feedback system which is sometimes called a 'closed' system, is influenced by its own past behaviour," *Principles of Systems* (Cambridge, Mass: Wright-Allen Press, 1968), p. 1-5.

2. In W.P. Brandon's words: "the way the word 'feedback' has insinuated itself in everyday discourse [stands in contrast] with the failure of other examples of scientific language to assume a similar place in ordinary speech." "The Concept of Feedback as an Explanation of Teleology," paper presented to the Annual Meeting of the Canadian Political Science Association, May 1978, p. 3.

3. "Strictly speaking . . . the concept of feedback applies only to information," but "the feedback loop will identify a set of processes, composed of information and related outputs and their consequences," D. Easton, *A Systems Analysis of Political Life* (New York: John Wiley, 1965), pp. 366 and 367; J.W. Forrester also defines a "Feedback Loop" as a "closed path connecting in sequence a decision [i.e., an output in Easton's vocabulary] . . . and information," J.W. Forrester, *Principles*, p. 1-7.

4. What follows is derived from S. Beer, *Brains of the Firm* 2nd Edition (New York: John Wiley, 1981) pp. 32–37.

5. *An Introduction to Cybernetics* (New York: John Wiley, 1956), p. 37.

6. What makes a situation acceptable or not is, of course, a most delicate question, which I have to set aside for the moment; but it should be made clear that, in the regulation of societal processes, ample attention should be given to the problem of the compatibility of technical norms of structural stability with the cultural norms of a society.

7. Actually, we need a program generating random values for the disturbances.

8. Note, however, that starting with the vector (VP,C,II) we obtain the equation:

$$(VP,C,II) \rightarrow (P,NC,II) \rightarrow (P,C,II) \rightarrow (A,NC,II) \rightarrow (A,C,II) \rightarrow (G,C,II)$$

in other words, the system settles at an acceptable level in this particular case.

9. For instance, Canada's federal system is composed of eleven governments; were they to communicate with each other all at the same time and in a reciprocal manner, the variety of the system would then be $11 \times 10 = 110$, but actually they may or may not communicate with each other at any given time, so that the variety is really 2^{110}. Note that the general formula is $V = 2^n$, where n is the number of bits of information in the situation at hand, see S. Beer, *Decision and Control* (New York: John Wiley, 1966), pp. 247–253.

10. R.W. Ashby, *Introduction*, p. 207.

11. For a good summary of Beer's ideas on this question see his *Designing Freedom* (Toronto: CBC Publications, 1974) from which most of what follows is taken.

12. See S. Beer, "The Surrogate World we Manage," in *Platform for Change* (New York: John Wiley, 1975), pp. 397–416.

13. *Designing Freedom*, p. 26.

14. Ibid., p. 29.

15. See "Fanfare for Freedom," in *Platform*, pp. 421–451.

16. Obviously, information screening in societal systems is a very complex process which Beer attacks here from only one angle, namely the impact of technology; there are many other aspects to this problem, such as the role played by groups which D. Easton calls the "gatekeepers" (see D. Easton, *A Systems Analysis*, Chapter 6). But here I am mostly concerned with what the cyberneticians themselves have to say on the subject of regulation.

17. See *The Brain of the Firm, in passim*.

18. See K. Deutsch, *The Nerves of Government* (New York: Free Press, 1966, 2nd ed.), pp. 185–192.

19. By asking "What is the effect of free public discussion, including freedom for unorthodox opinions, upon the predictive efficiency of a political decision system?" (ibid., p. 190), Deutsch seems to imply that the concept of lead gives us a rational basis for the defense of political freedom. There again we see that cybernetic theory should not be confused with a prescription for a 1984-type of automated society; my only criticism is that Deutsch should have elaborated that point much more extensively.

20. See R. Landry and V. Lemieux, "L'analyse cybernétique des politiques gouvernmentales," *Canadian Journal of Political Science*, vol. XI, No. 3 (September 1978), p. 534, also R. Landry, "Elements pour une nouvelle génération d'analyse systèmique en science politique," paper presented at the Annual Meeting of the Canadian Political Science Association, Fredericton, 1977.

21. W.R. Ashby, "Adaptation in the Multistable System," (from *Design*), in F.E. Emery (ed.), *Systems Thinking* (Penguin Books, 1969), p. 230.

22. For a summary of H. Simon's ideas on this question and their refutation with the help of a simulation model, see D. Metlay, "On Studying the Future

Behavior of Complex Systems," in T.D. LaPorte (ed.), *Organized Social Complexity* (Princeton, N.J.: Princeton University Press, 1975), pp. 220–255.

23. See ibid., *in passim*.

24. This model is borrowed without qualifying assumption from classical physics; and it can only be applied to animal or human behavior if the postulate of reversibility, inherent in the mechanical paradigm, obtains, which is rather hard to accept.

25. The author have specifically in mind the Hullian learning theory, see "Purpose and Learning Theory," in W. Buckley (ed.), *Modern Systems Research for the Behavioral Scientist* (Chicago: Aldine, 1968), pp. 250–255.

26. Deutsch alludes here to G. Bateson's "deutero learning"; see *Nerves*, p. 169.

27. C.H. Waddington has given a clear and brief account of his homeorhetic model in *Tools for Thought* (New York: Basic Books, 1977), pp. 103–116; see also his "Concluding Remarks" to E. Jantsch and C.H. Waddington (eds.), *Evolution and Consciousness* (Reading, Mass.: Addison-Wesley, 1976), pp. 243–249, and his *Strategy of the Genes* (London: Allen and Unwin, 1957).

28. J. Piaget, *Biologie et connaissance* (Paris: Gallimard, 1967), p. 45 (my translation).

29. *Tools*, p. 110.

30. See R. Thom, *Structural Stability and Morphogenesis* (Reading, Mass.: Addison-Wesley, 1976); C.H. Waddington was one of the first to discern the practical import of Thom's theory.

31. *Tools*, p. 113.

32. See *op. cit.*, pp. 210–211.

33. F.G. Varela, "The Grounds for a Closed Logic," *BCL Report*, No. 35 (Urbana, Ill.: Biological Computer Laboratory of the University of Illinois, 1975), pp. 7–8. Note also that the elaboration of this logic of closedness should help us to conceptualize the creative role played by the observer-as-a-subject in the apprehension of his world—in our case, the societal world and its self-regulation; this crucial point is discussed in Chapter 6.

34. E. Morin, *La méthode*, p. 212.

35. See ibid., pp. 193 and 214.

36. This point has also been made by Yves Barel, who sees reproduction as a generic phenomenon of which biological and social reproduction are only particular cases. For him, the logical link between production and reproduction rests in the concept of differentiation; the production of differences leads to the closure of the system and the reproduction of these differences. This is, however, only one side of the question, the other being that differentiation also threatens the invariance of the reproductive process. See Y. Barel, *La reproduction sociale* (Paris: Anthropos, 1973), pp. 31–62.

37. See Howard Gardner, *The Mind's New Science* (New York: Basic Books, 1985); D.R. Hofstadter and D.C. Dennett, *The Mind's I: Fantasies and Reflections on Self and Soul* (Harmondsworth: Penguin Books, 1981); and John Haugeland (ed.), *Mind Design* (Cambridge, Mass.: MIT Press, 1981).

38. Etzioni analyzes the concept of societal consciousness, including self consciousness (see *The Active Society*, pp. 223–248), but he does so in socio-psychological terms. I do not imply that he is wrong in doing so but rather that he misses a crucial step, namely the epistemological significance of the hypothesis that something like a collective self is evidenced in social phenomena.

39. See *La méthode*, p. 239.

40. See *Nerves*, Chapter 13; interestingly, Deutsch locates the "self" of an organization in the points where "a stream of data recalled from memory" is fed back into "a stream of decisions regarding current behavior" (ibid., p. 220).

41. Ibid., p. 221.

42. R.W. Chadwick, "Steps Toward a Probabilistic Systems Theory of Political Behavior, With Special Reference to Integration Theory," in H.R. Alker, Jr., et al., (eds.), *Mathematical Approaches to Politics* (San Francisco: Jossey Bass, 1973), p. 69.

43. "Specific support" is a response on the part of the members of the political system to identifiable benefits, whereas "diffuse support" is unconditional, see *A Systems Analysis*, Part V.

44. Ibid., p. 371.

45. *Nerves*, p. 91; Deutsch aptly writes "restoring" insofar as liberalism has pushed the notion of a common good out of politics and into economics, and positivism and banished it altogether from social thought.

46. Ibid., p. 234.

47. P. Birnbaum, *La fin du politique* (Paris: Seuil, 1975), p. 128 (my translation); see also W.P. Brandon, "Feedback," pp. 21–23.

48. A. Rosenblueth, N. Wiener, and J. Bigelow, "Behavior, Purpose, and Teleology," in W. Buckley (ed.), *Modern Systems Research*, p. 225.

49. The whole debate between N. Wiener et al., and R. Taylor is to be found in Chapters 27 to 30 of W. Buckley (ed.), ibid.; see also H. Jonas, *The Phenomenon of Life* (New York: Harper and Row, 1966), pp. 108–244.

50. I concede to Brandon that the use of the terms "goal" or "target" is "revealing" of a reluctance to deal with the whole reality of purpose, see "Feedback," pp. 9–10.

51. See W.R. Ashby, "Setting Goals in Cybernetic Systems," in H.W. Robinson and D.E. Knight (eds.), *Cybernetics, Artificial Intelligence, and Ecology* (New York: Spartan Books, 1972), pp. 33–44.

52. I do not imply that this school of thought denies the existence of something called "values," but rather that it treats them as attributes of political phenomena similar in all respects to any other attribute, having no particular meaning for the observer himself. Typical of this attitude are A. Kaplan's methodological instructions for the study of values in *The Conduct of Inquiry* (Scranton: Chandler, 1964), Chapter X and E. Nagel's critique of M. Weber's writing on this very subject in *The Structure of Science* (New York: Harcourt, Brace and Co., 1961), pp. 481–483.

53. See Sheldon Wolin, *Politics and Vision* (Boston: Little, Brown and Co., 1960).

54. G. Bateson, *Steps to an Ecology of Mind* (New York: Ballantine Books, 1972), pp. 315–320.

55. Landry and Lemieux take the coercive aspects of control into account, even faulting most systems theorists for reasoning "as if information flowed [within the system] without manipulation or any other constraint" ("L'analyse cybernétique," p. 541), and yet fail to acknowledge the normative bias thereby introduced into cybernetics.

56. *Brain of the Firm*, p. 52.

57. R.M. Tostevin, "Living Without the Mastery of Nature: Beyond the Spell of Enlightenment," (Toronto: unpublished dissertation, York University, 1977), pp. 244–246.

58. See F. Varela, *Principles of Biological Autonomy* (New York: Elsevier, 1979), Preface.

59. See W.P. Brandon, "Feedback," p. 4, and R. Tostevin, "Living Without the Mastery of Nature," pp. 342–370, where the latter author opposes Aristotle's ethics (re: *The Nichomachean Ethics*) to the cybernetic model of learning, pleading for "prudence" against the obstinate determination to experiment with the unknown.

60. Note that, at the level of practice, the cybernetic approach, in which "errors" play an important role, may turn out to be a good antidote to dogmatism and simplistic attitudes. A planner fully aware of the organized complexity of societal processes, and reasoning in terms of mutual causal interactions, is more likely to recommend the adoption of what LaPorte calls "reversible" policies, which can be modified in the event of "counterintuitive" societal developments; see T.D. LaPorte, "Complexity and Uncertainty: Challenge to Action," in T.D. LaPorte (ed.), *Complexity*, pp. 332–356.

61. E. Jantsch, "Evolution: Self-Realization Through Self-Transcendence," in E. Jantsch and C.H. Waddington (eds.), *Evolution and Consciousness: Human Systems in Transition* (Reading, Mass.: Addison-Wesley, 1976), p. 63.

3

Self-Organizing Complexity

If the first cybernetics did not make good on its original insight concerning the emerging autonomy of self-regulated systems, emphasizing, as it did, control rather than autonomy, the second cybernetics takes up that challenge. It is a meta-theory, a "cybernetics of cybernetics" (H. von Foerster) which formally accounts for a meta-process: the organization of the generative process that creates and destroys the concrete and yet transient structures composing complex systems. In a sense, Artificial Intelligence (AI) research has laid down the foundation for a third generation of cybernetic models. However, AI research is so diversified that it is difficult to identify a single typical paradigm. This is why I limit myself hereafter to a discussion of the theory of self-organization.

The basic concepts of the theory of self-organizing systems are defined in the first section; their technical implications are worked out in the second section in which von Foerster's "order from noise" principle is analyzed; the last section attempts to make sense of the immense complexity of societal disorder by constructing a three dimensional analytical grid for the investigation of the various ways in which societal structuration processes unfold. The epistemological justification for this entire exercise is that it is proposed as a feasible way of moving beyond the strictures of either the mechanistic or the organismic world views.

Autopoiesis: The Regeneration of the Self

> A living system . . . possesses and combines to an exceptional degree systemic complexity, the complexity of open systems and cybernetic complexity . . . It might be assumed that, having said this, we have adequately described the complexity of living things, and that it would suffice to bring in systems theory and cybernetics . . . [Yet] the complexity peculiar to living things,

> while it comprehends these orders of complexity, is of another . . . quality, and answers to a different organizing principle.[1]

To this E. Morin adds that "*Homo sapiens* raises the problem of hyper-complexity."[2] At stake here is our capacity to conceptualize systemic processes that go beyond structural adaptation. Biologists and social scientists have in common that their subject matters confront them continually with paradoxes. For instance, living organisms appear to be characterized as much by their instability, evidenced by the fragility of their component parts (e.g., the high mortality rate of individual cells), than by their homeostatic behavior. What biological regulators regulate is not entirely clear, and the same argument applies with even more pertinence to societal regulators.

The hierarchical structuration of complex systems is a widely accepted notion.[3] By "system" we must always mean a totality whose unity is consistent with a high degree of internal diversity. It follows from the critique of cybernetics articulated above that the representation of political processes as a cybernated system can be relevant only if it opens, at some point, on a still higher level of complexity. And here the possibility of a breakthrough depends on the merits of a radical distinction between the concepts of "organization" and "structure." For L. Loefgren, "structure within model theory is a set endowed with maximal set of relations with respect to the predicated symbols of the theory to be interpreted in the structure . . . hence [it] may be looked upon as a concrete object."[4]

Similarly, H. Maturana and F. Varela define structure as "the actual relations which hold between the components . . . [and] integrate concrete machines in a given space."[5] Pieces of legislation (electoral laws, taxation laws, etc.) and less formal conventions specify the rules of the political game, and as such enter into the definition of societal structures. Sometimes these rules can be expressed as a set of equations, for instance within the framework of game theory.

Structure is, of course, a widely used term, notably in mathematics, anthropology, linguistics, and sociology. The latter three disciplines have indeed been profoundly influenced by structuralism, a meta-theory which attempts to explicate symbolic systems in their totality.[6] However, political scientists, who use the term frequently as a synonym for institutional arrangement, do not attach any clear theoretical meaning to it. In mathematics, a structure is defined as a set of elements together with rules of transformation (i.e., operations) applicable to these elements. For example, a given vector space constitutes a particular structure contained in the all-inclusive set of vector spaces. This definition is consistent with those of Loefgren or of Maturana and Varela. In the social sciences, however, the distinction between a certain structure and its generic form is not always clearly maintained. Space is lacking here to undertake a critical review of structuralist ideas, but I would like to venture that the debate over the "reality" of structures stems from a failure to consider seriously this semantic slippage. To avoid falling into the same trap, it is necessary to distinguish between structure and "organization." In F. Varela's words: "In contradistinction to structure, the actual relations holding between the components of a given system, organization refers to classes of relations which can be realized by several structures."[7]

The organizing principle of a complex adaptive system produces and transforms that system while maintaining its unity through time. According to E. Morin, there is a reciprocity between the concepts of organization, system and interrelation: "A system is the phenomenal and global character acquired by interrelations the ordering of which forms the organization of the system."[8] This conceptual trinity challenges us "to perceive, conceive and reason in an organizational way what surrounds us, which we call reality."[9] What Morin means is that when we look at any entity, be it atoms, biological organisms, or societal systems, we should always be aware of the difference between their enduring form, or organization, and its transient realization into concrete objects, or structure. Now, the organization of a system needs not to be thought of in Platonic terms as an immaterial Idea, but rather as an actuality emerging on a different time scale. D. Easton follows a similar line of reasoning in distinguishing between the regime, or constitutional order, and the political community which manifests a higher degree of resilience.[10]

We come here to a critical juncture. It is now time to introduce the concept of recursion. The inclusion or "nesting" of an operation within itself is the defining characteristic of a recursive structure. A computer program, for example, often uses data generated by one or more sub-programs that may themselves include one or more tertiary programs and so on; when the sub-program consists of a replication of the master program which is thereby included within itself, a recursion occurs. This procedure does not lead to an infinite regress because each step is hierarchically related to the next one. Therefore, it is always possible to determine the level of recursion at which the program is operating and from there to return to the execution of the master program.[11]

It follows that a recursive structure refers to classes of relations, each one of which can be realized by several (sub-)structures and, therefore, can take many distinct concrete forms. Does this mean that a recursive structure is necessarily an organization as defined above by F. Varela? It seems to be at best an approximation of it, insofar as the kind of systems in which Varela is interested are not simply recursive (i.e., programmed by an *external* agent) but actually "autopoietic."

> An autopoietic machine is a machine organized (defined as a unity) as a network of processes of production (transformation and destruction) of com-

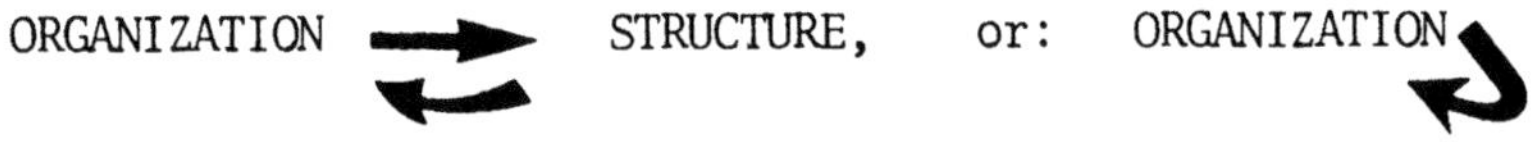

Figure 3.1

> ponents that produces the components which: (i) through their interactions and transformations continuously regenerate and realize the network of process (relations) that produced them; and (ii) constitute it (the machine) as a concrete unity in the space in which they (the components) exist by specifying the topological domain of its realization as such a network.[12]

Non-autopoietic systems are said to be "allopoietic." Again, what a system is open or closed to can be taken as an indicator of its comparative complexity. It appears thus that autopoietic systems are more complex than allopoietic ones since by definition the former are open at the level of their goals and closed only at that of their products, whereas the latter are closed at the level of their program and only open at that of their product.[13] The fundamental characteristic of an autopoietic system is its *autonomy*. It is not programmed to respond to its environment by correcting its outputs, but it maintains its integrity by modifying its own internal procedures.

Life is autopoietic. The simplest manifestations of self-renewal occur in cells,[14] but the regenerative capacity of living systems is evident at all levels and has always fascinated the human mind. What is continually in the process of being organized in a living system is the organization of the living system itself (see Fig. 3.1).[15] If living systems are self-organizing and self-renewing, does it mean that politically organized communities are also? The transition from biological to societal realities always remains problematic. In fact, for F. Varela, autopoiesis is primarily a model for understanding physical processes.[16] Other authors, however, insist that autopoiesis is the concrete manifestation of a more universal class of phenomena which display a tendency toward self-organization; a wide range of non-organic, living, and social systems, in their own ways, reproduce themselves by evolving various types of organizational closure.[17] It is in this sense that I understand the import of the self-organization paradigm for political science. It is neither strictly limited to the analysis of biological organisms nor is it a grand theory that explains everything about all complex systems. It is a way of posing problems, a point of departure, i.e., to what extent is this or that particular set of relations characterized by its autonomy and its capacity to act upon itself.

However, self-organization cannot be interpreted as some kind of perpetual motion; there are certain conditions to its realization. R. Ashby has indeed proven that self-organization, in the strictest sense of a system that would find in itself all the resources required for its development, is an impossibility.[18] Such a system always exists in, and at the expense of, a rich environment. Pasteur's experimental disproof of the spontaneous generation of life provides us with a good illustration of this point. Even the organization of abstract conceptual systems, or "noogenesis" as some authors prefer to say, exacts its price; indeed Henri Bergson tirelessly proclaimed that intellectual creation supposes duration, and thus draws on the scarcest of all resources, namely time measured against one's life-span.

Because self-organization results in the creation of order or, more exactly, of a secondary order (an "order of process," to use Laszlo's phrase), patterns of self-organization can only be understood in relation to all the processes they act upon, both inside and outside the boundaries of a complex system. I argue in the next section that these processes are contradictory and yet complementary.

The Creative Function of Disorder

While scientific thinking has generally emphasized quantitative measurement, the profound significance of the paradigm of self-organization lies precisely in that it is based on the qualitative notion of order. It has therefore an almost universal range of applications.[19] To formalize the notion of order, as the paradigm of self-organization invites us to do, is not a vain metaphysical exercise. Order in the polis is both an end and a means to politics—to define politics in terms of conflicts, or disorder, is nothing but a mirror image of this tautology. Thus, even if there is no direct correspondence between political theory and the formal models of self-organization presented hereafter, the latter may nevertheless serve as catalysts for a new synthesis of the fundamental, but elusive, notion of political order.

In self-organizing systems both negative and positive feedback loops interact. The former channel "deviation-counteracting" patterns of information, and the latter channel "deviation-amplifying" patterns that bring about situations in which "[a] small initial deviation, which is within the range of high probability, may develop into a deviation of very low probability."[20] Any move away from its structural norms represents a real, or at least potential, threat to the system. Yet some of these threats may turn out to open unforeseen paths of development. M. Maruyama illustrates this insightful conjecture with a series of examples (among which the role played by mutations in phylogenesis is perhaps the most suggestive), but does not formally establish its validity. However, von Foerster's demonstration of the creative function of "noise" in communications can be invoked in support of Maruyama's hypothesis and requires careful exposition.

In E. Schrödinger's words, orderly events result from either the performance of complex structures feeding upon negative entropy, a principle which he called "order from order," or from "the statistical mechanism which produces order from disorder."[21] Homeostasis is the best example of the former principle, while stochastic processes are so common in the modern scientific world-view (from the Brownian movement of molecules to the fluctuations of public opinion) that Schrödinger's phrase needs not to be elaborated upon. However, the so-called Law of Large Numbers, to which Schrödinger is alluding, does not account for all the effects of random variations. In particular, H. von Foerster argues that self-organizing systems are capable of sorting out, in a flow of random inputs, the information required to produce new structures; von Foerster calls this "order from noise," a principle which he claims has nothing to do with Schrödinger's "order from disorder" principle.[22]

In a formal language, self-organization is equivalent to an increase in the "redundancy" of a system, according to von Foerster. Let H be the entropy of this system—considered as an information source—and H_m the maximum value that H can take; by definition the redundancy R is a function of the ratio of H to H_m: $R=1-H/H_m$, and $dR/dt= -[H_m(dH/dt)-H(dH_m/dt)]H_m^2$. Von Foerster's definition of self-organization requires that $dR/dt > 0$, a condition which can only be realized if $H(dH_m/dt) > H_m(dH/dt)$.[23] Two necessary and sufficient conditions for the validation of this equation are either that the actual amount of entropy in the system be restored after having initially decreased ($dH/dt < 0$) or that the maximum amount of potential entropy increases ($dH_m/dt > 0$), and in this sense noise plays its part in the process of self-organization. It may be objected that the entropy of a social system is generally unknown, and that consequently the above demonstration is of very little relevance to political scientists. However, Stephen Coleman argues that the objective measure of the unpredictability of political systems should be regarded as one of three (potentially four) fundamental measures upon which a science of politics can be built; this author goes on to suggest that elections are social situations fitting all the requirements for the measurement of entropy, and that this approach leads to the formulation of an original hypothesis about the relationship between voters turn out and the expression of partisan preferences.[24]

To equate order with redundancy, as von Foerster does, is not only consistent with information theory but also justified by our knowledge of the physiology of living systems. Nevertheless, such a definition becomes untenable when taken too literally since it would tend to reduce complexity to predictability. But only simple phenomena are completely predictable. It is equally true that a system whose behavior would be totally unpredictable could not be self-organizing. Henri Atlan takes these two opposite constraints into account in his own model which offers therefore a more dialectical interpretation of the "order from noise" principle. He argues against the transposition of "the notion of information transmitted through a communication channel to that of information contained in an organized system"; in the latter case, "[noise] does not mean a dissipated quantity of information but, on the contrary, an increase in the variety of the whole system, that is to say a decrease in its redundancy."[25] However, no system can continue to exist below a certain threshold of redundancy, thus noise is both a destructive and a constructive factor ("destructive ambiguity" and "autonomy-ambiguity" in Atlan's words).

In Atlan's model H stands for the quantity of information contained in a given system and H_m for the maximum quantity of information that this system could communicate to an observer about itself if no symbol in the message was redundant. By definition we have $R=1 - H/H_m$, thus $H=H_m(1 - R)$, and $dH/dt = - H_m dR/dt + (1 - R)dH_m/dt$, or $dH/dt = -H_m f_1(t) + (1 - R)f_2(t)$. Note that $f_1(t)$ expresses the effect of "autonomy-ambiguity," or the creative function of internal instability, and $f_2(t)$ records the extent of disorganization caused by "destructive ambiguity." For Atlan dH/dt should be positive in the case of a self-organizing system, since an increase in H means that the system becomes able to respond to more demands from its environment (Ashby's law of requisite variety); this implies that $f_1(t)$ should be negative, contrary to what von Foerster assumes.[26] Of course, Atlan's model rests on the hypothesis that the self-organizing system possesses an initial level of redundancy high enough for $f_1(t)$ to remain negative for a sufficiently long period of time.

These two models could precisely be linked at the crucial point where redundancy, beyond a certain threshold value, becomes a source of rigidity. In any case the superior elegance of Atlan's model resides in that it illustrates quite clearly how the same internal disturbances affecting in opposite ways $f_1(t)$—which tends to become negative—and $f_2(t)$—which tends to become positive—contribute, at first, to the growth of the system—when $f_1(t)$ prevails—and then to its degeneration—when $f_2(t)$ prevails.

To add some substance to these admittedly rather abstract developments, we can draw a parallel between the "order from noise" principle and Prigogine's "order from fluctuations" principle which, by contrast, is firmly grounded in experimental research. Nonequilibrium systems—or "dissipative structures," as Prigogine calls them—do not, under all circumstances, drift irreversibly toward the lowest rate of entropy production.[27] This had already been pointed out by Bertalanffy, but Prigogine has gone further in the sense that he has spelled out explicitly the thermodynamic laws (expressed in stochastic terms) associated with such processes. In brief, as E. Jantsch explains:

> If systems of any kind are in a sufficiently non-equilibrium state, have many degrees of freedom and are partially open to the inflow of energy (information) and/or matter, the ensuing instabilities do not lead to random behavior (even

> if the initiating fluctuations and the mutations as such are random); instead, they tend to drive the system to a new dynamic regime which corresponds to a new state of complexity.[28]

This is a generalization of what Prigogine established as being true of certain physical systems (e.g., vortices occurring in fluids) and biological organisms (e.g., the process of embryonic development). He has suggested also that the "order through fluctuations" principle could also be applied to the study of population dynamics in animal societies and even to the modeling of societal systems.[29] At the very least, Prigogine and Stengers' suggestion that social systems "are highly sensitive to fluctuations" is worth investigating.[30]

It is not difficult to find examples of overlapping destructuration and restructuration on many planes. For example, the recent experimental discovery of cases of "learning by dislearning" in children brings an interesting nuance to the Piagetian theory of cognitive development through sequential structural changes.[31] Perhaps the clearest manifestation of the creative function of disorder is painfully close to us: as each generation enriches the culture that nurtured it, another one passes away.

There is clearly a danger here; if the "order through fluctuation" principle helps us to understand many complex dynamic processes, it may be tempting to claim that it could unlock universal complexity. Peter Corning has accused E. Jantsch of having done just that when he wrote that the thermodynamics of systems far from equilibrium "governs" the evolution of all complex systems.[32] Certainly, the problem is real and Jantsch may well have been carried away in this particular instance. However, Corning's criticism misses the point. In order to prove Jantsch wrong, he argues that "If there has been order through fluctuation, there has also been disorder through fluctuation," pointing to the countless examples of evolutionary dead-ends.[33] This is nothing more than a rhetorical attack, for neither Prigogine nor Jantsch ever denied that internal disorder is capable of destroying whatever organization there exists in a system experiencing a crisis. "Order through fluctuation" is not a rigorous law that always, and under all circumstances, results in a beneficial increase in the internal complexity of every physical, biological or social system.

The Dimensions of Societal Disorder

There are obvious analogies between the dual principles of order from order and order from noise, on the one hand, and political life on the other.[34] Modern political thought appears in many ways like a quest for the key that would make sense of the very "noisy" cacophony of individual and group aspirations. The classical political economists' idea that a collective good is achieved by the competition of narrow individual interests is perhaps no less dialectical than historical materialism. But the Marxist school is, of course, much more explicitly concerned with the ambiguity of the relationship between social order and disorder. Thus, it would be naive to try to develop a generalization of the above mentioned cybernetic principles, at the level of societal dynamics, without taking Marxist theory into consideration. I do not want to rediscover, as it were, the notion of historical contradiction. And yet, I will argue that a farreaching revision of Marx's original conceptions is required to comprehend the apparently worsening internal tensions sustained by the economic and political institutions of "post-modern" societies. The renewed interest in, and elaboration of, the notion of crisis in social scientific research, prompted by a cohort of authors beginning with J. Habermas, renders this venture possible. I believe that it is a promising research agenda. At the same time, and as a note of caution, we should heed N. Luhmann's witty remark that social scientists are prone to call crisis whatever development they observe which their paradigm had not led them to anticipate.[35]

How and to what extent do systemic disorders play a functionally similar role in both human and non-human systems? We are faced here with two problems. First, the concept of self-organization begs the question of knowing when a system capable of regenerating its own structures ceases to be itself altogether.[36] This problem is inherent in systems theory. (As A. Rapoport puts it, "the cardinal principle of 'systemness' [is] constancy amid change.")[37] But it seems to be more serious in a social context. The second methodological obstacle concerns the proper level of analysis. What is being threatened in a socio-political system in a position far from equilibrium: the concrete regulators or the ideational representations through which they operate, or both?

Identity Through Time

The question as to whether the universe is one and unchanging, or diverse and turbulent has been a perennial preoccupation since the Presocratic era. My intention here is not to "solve," once and for all, the problem of the identity through time of complex systems, but rather to formulate it in such a way that it should not form an a priori insuperable obstacle.

It is a common misconception to assume that the spatial and temporal boundaries of man-made systems, and of political communities in particular, are more difficult to delineate than those of concrete systems. This is not necessarily the case. Indeed, some concrete systems are very puzzling; for instance, there is a great deal of speculation as to what remains in a "black hole" of the original energy field of a star. Even the birth and death of biological organisms are becoming equivocal events with the progress of embriology and geriatrics (e.g., the debates over the legalization of abortion on demand, or on the fate of comatose patients). And the search for the origins of *Homo sapiens* gets more frustrating with each new discovery. The fact that the natural sciences seem to move towards an evolutionary, time-bound perspective, despite the imprecision that such a viewpoint introduces in the definition of their objects of inquiry, does not justify the application of a similar perspective to political phenomena. It tends, however, to shift the burden of proof onto those who claim that such an approach is not "scientific." Probability theory is a useful tool for analyzing changes in the distribution of any set of parameters in a system. The application of these methods to the modeling of various types of stochastic processes occurring in nature is common. However, most socio-political systems are such that the very definition of the constant and varying patterns is problematic and depends, in part, on subjective factors.[38]

An interesting strategy for coming to terms with the identity of self-organizing systems consists in distinguishing, as G. Pask has done, between "tacit" and "language oriented" systems.[39] The former are those systems for which the observer prescribes a goal. When they fail to realize that goal or limited set of goals—or when it loses its significance, such systems cease to exist in the universe constructed by the observer. "Language oriented" systems, on the other hand, state their own goal(s) and are capable of specifying how and when they may change course. In other words, they can enter into a dialogue with the observer who, in the process, loses some of his objectivity and becomes an interpreter; his aim is to understand rather than to prescribe and predict, and his method is hermeneutical rather than positivistic. J. Habermas offers a similar argument:

> . . . social systems too have identities and can lose them; historians are capable of differentiating between revolutionary changes of a state or the downfall of an empire and mere structural alterations. In doing so, they refer to the interpretation that members of a system use in identifying one another as belonging to the same group.

But he adds that "[a] society does not plunge into a crisis when, and only when, its members so identify the situation."[40] The balancing of the objective and subjective dimensions of a system's identity is probably the most difficult task facing the theoretician of systems.

At the very least, to speak of a dialogue, that is to say a meaningful communication between thinking subjects, is to suppose that each subject knows the rules of the debate, as it were. On the agora of Greek city-states political and philosophical debates were settled on the basis of some simple rules of logic—sometimes twisted by the not so simple art of Socratic dialectic. But on the modern agora too many voices, speaking too many languages, clamour in confusion. We therefore need a "cybernetics of cybernetics," as H. von Foerster put it, that is to say a "cybernetics of observing systems," in order to comprehend the intricate network linking the observer(s), the social systems he (they) interpellates, and the languages they use.[41] Such a network is likely to be self-organizing, in which case each one of its poles should be sufficiently redundant to take advantage of the "noise" that will inevitably accompany the messages exchanged among them.[42] In other words, the dialectic of order from order and order from noise is replicated at the level of cognitive development. Thus knowledge is active; it creates and breaks societal structures and shapes the languages that support their existence. The identity of social systems is not something hidden that social scientists can unveil through patient research, although there is a need for that kind of detached analysis, but rather is something in the determination of which they (and natural scientists alike) play a part.

The categories "language oriented" or "tacit" systems, and human or non-human systems tend to overlap. However, they are not equivalent. Some social organizations can, for all practical purposes, be regarded as tacit, insofar as they derive their rationality from external sources (e.g., highly automated or capital intensive production systems).[43] (Note also that policy analysts often define a policy field as if it was composed only of things (e.g., hospital beds, medical facilities, etc.), whereas politicians are more sensitive to the human complexity of policy choices because they come into contact with a multiplicity of pressure groups.) Inversely, animals seem capable of imparting to the observer a more or less accurate realization of their own purpose, namely their urge to survive either individually or collectively. This is sometimes expressed quasi-symbolically and even, in the case of higher species, consciously realized to some extent. But there is no doubt about the nature of political systems. Their symbolic richness brings us precisely to our second dilemma, that of the appropriate level of analysis.

Crises Within Crises: Experience, Consciousness, and Self-Consciousness

We have seen that, in the simple case where the receiver does not need a meta-code to interpret the message (I would call this a "horizontal" communication), the process of self-organization blurs the dividing line between information and noise. This, as von Foerster and Atlan have shown, can be easily formalized. But the semantic richness of the concept of information is not exhausted by these formal models.[44] Living systems develop in a multidimensional space, and this is even truer of societal systems. To illustrate the complexity of political, economic, and ideological disorder, I hereafter propose a three-tiered scale of societal crises which takes both the objective and subjective aspects of political life into account.

The first type refers to crises that result from the empirically observable breakdown of any given societal arrangement upon which the very conditions of our subsistence may depend; the "stagflation" and the depression of the late 1970s and early 1980s is a case in point. What I would call crises of societal consciousness, on the other hand, occur when our private neuroses are expressed through institutional reform (or revolution) and thereby partially revealed to by our practical consciousness. Eventually, cultural crises threaten the legitimacy of the social and political order; the outcome of such crises is a more or less profound change in the ways in which political life is theorized and in the ways in which political debates are articulated. Of course, these types of crises should not be understood as being unrelated or mutually exclusive. However, for the sake of clarity, I intend now to discuss the nature of these crises as if they were analytically distinguishable.

The notion of crisis has a long history. In Greek, "*krisein*" meant interpretation (of omen) or choice (of the sacrificial victims). Hippocratus introduced the term into the medical vocabulary, and we owe to J. Habermas a penetrating exegesis of its meaning in that context:

> [crisis] refers to the phase of an illness in which it is decided whether or not the organism's self-healing powers are sufficient for recovery. The critical process, the illness, appears as something objective. A contagious disease, for example, is contracted through *external* influences on the organism; and the deviation from . . . the normal state [of the organism] can be observed and measured with the aid of empirical parameters. The patient's consciousness plays no role in this; how he feels . . . is at most a symptom of a process that he himself can scarcely influence at all. Nevertheless, we would not speak of a crisis . . . if it were only a matter of an objective process viewed from the outside, if the patient were not also subjectively involved in this process. The crisis cannot be separated from the viewpoint of the one who is undergoing it—the patient experiences his powerlessness vis-a-vis the objectivity of the illness only because he is a subject . . . deprived of the possibility of being . . . in full possession of his powers.[45]

This duality of meaning did not escape Marx who gave to the notion of crisis a most central place in his theoretical apparatus. The kind of crisis envisaged by Marx is a catastrophic disruption of the productive process. As a result of the unequal distribution of profits, an imbalance between the output of commodities and the purchasing power of the economic agents develops over the years until it reaches a breaking point.[46] But, following a logic strikingly similar to that of Prigogine's thermodynamic model, Marx explains that the crises of overproduction produce the conditions of their own solution:

> From time to time the conflict of antagonistic agencies finds vent in crises. The crises are always but momentary and forcible solutions to the existing contradictions. They are violent eruptions which for a time restore the disturbed equilibrium.[47]

In particular, each crisis causes a devaluation of capital which allows the development of yet another cycle of accumulation. At the same time, it reveals the hidden contradictions of the system, thus preparing the ultimate overthrow of capitalism. An economic crisis leads to a political one. It is here that the dialectical interplay of objective and subjective factors takes place.

Moreover, we find in the Marxist discourse the analysis of a more global crisis. The so-called permanent crisis of capitalism subsumes and surpasses those violent, but momentary and in a sense healthy, upheavals. This concept is incipient in Marx's work, but is more clearly worked out in Lenin's *Imperialism: The Last Stage of Capitalism*, and is still a subject of much debate in contemporary Marxism. Like overproduction crises, the permanent crisis furthers an awareness of the antagonistic tendencies of a world-wide explorative process, but the permanent crisis does not simultaneously bring about its own solution. It is supposedly fatal to capitalism and its ultimate outcome remains uncertain. The temptation is great to reject this analysis on the grounds that it is derived from a quasi-eschatological rather than scientific understanding of social realities. Indeed, it is based on an act of faith, but it must be conceeded that the transition from the short term to the long run projects the observer into an unfamiliar world, forcing him to build new images of his condition. Whether one does or does not share these representations is another matter. At any rate, it is probable that the effects of positive feedback loops, which are highly aleatory, become more perceptible as one moves along the time scale.[48]

Setting aside the problem of the ultimate fate of capitalism, we return now to Marx's analysis of overproduction crises and of their social and political consequences. The reality of cyclical economic crises is unquestioned; in fact, neo-classical economists regard them also as useful means of clearing the economic climate, even if they do not attribute their origin to the same causes. There is indeed much to be argued about the relationship between capital accumulation and the variations of the surplus value, a most specious concept. In particular, this theory escapes any kind of empirical validation. But my purpose here is not to assess the overall validity of Marx's political economy, it is rather to establish whether or not there exists a social theory that could provide a locus for the application of the order from noise principle. And, at first sight, it seems that Marx's analysis of economic crises meets this requirement. I intend to show now that it does not.

The crucial question concerns the centrality of economic structural changes in the process of societal self-organization. It is, of course, a Marxist dogma that the infrastructure (viz. the productive forces) determines, in "the last analysis," the political and ideological superstructures. But to posit the complete autonomy of the political sphere would be equally dogmatic and no more convincing. If nature is as turbulent as modern science seems to imply, it seems only logical to assume that man will experience an echo, if not an amplified resonance, of this turmoil in his attempt to control nature. And the shock would be felt most strongly in the economic sphere, but would certainly reverberate throughout the societal body. Note, however, that I do not suggest that the origin of all societal disorders is to be found invariably in economic disturbances, nor even in man's dependence on natural resources; the point is simply that such a chain of events is a common occurrence. What is problematic from a methodological standpoint is the representation of the trajectory of this impulse, as it were. Marx's allegation that economic disorders almost immediately give rise to political demands and acts, resulting in the creation of a new political order (i.e., a redefinition of the balance of power between the classes), as well as ideological justifications of the latter, did not make such sense in his time, and make even less sense now. Even if such a characterization of Marx's dialectical analysis is crude, it remains the case that the distinction between the objective and subjective dimensions of political conflicts, which underlies Marx's critique, rests on an inadmissible reification of the notion of social disorder. The informational, and thus essentially ambiguous, content of the act of production itself has become obvious in the post-industrial era, but was always inherent in that process.

One cannot, however, treat information as a one-dimensional concept, and here I refer again to J. Attali's classification.[49] In fact, this author has developed an interesting reformulation of the Marxist schemata in light of the order from noise principle. For him, "the space of prices and value," that is to say the theoretical locus of Marx's notion of crisis, "is only the expression of the first two levels of information," namely cybernetic signals and semantic information.[50] Structural changes at these levels result in a re-organization of the corresponding codes which react upon political as well as economic transactions. It might be recalled that in Attali's model the "cybernetic code" determines the distribution and reproduction of valued resources, and thus entails a political choice as to what kind of role is played by individual producers, corporations, state bureaucracies, etc. (i.e., "free" competition, decentralized or centralized planning, etc.). The semantic code again subsumes scientific knowledge and cultural productions. What makes Attali's analysis so original is that he sees the origin of the "noise" affecting the "space of prices and value" in random "aggressions" perpetrated on it by symbolic images. The fundamental crisis of advanced capitalism, he goes on to conclude, manifests itself in the desperate but impossible efforts deployed by the ruling class to contain cultural innovations and "imaginary information" within the limits of commodity exchanges. Here he seems to me to overstate his case, but regardless of the value of his diagnosis, Attali's suggestion that the symbolic richess of "language oriented systems" is their most important source of "noise" deserves attention.

Attali does not himself elaborate what he means by imaginary information. To further explore this dimension (which closely corresponds to the notion of consciousness in my trilogy) we must turn to the variously successful attempts to use psychoanalysis (or other methods of psychological analysis) in social theory. A complete review of this by now vast field would obviously exceed the limits of our topic, but it is hard to ignore Marc Guillaume's *Eloge du désordre*. This author arrives at the following hypothesis without a first-hand knowledge of von Foerster or Atlan's research, but through a reading of Lacan, Foucault, and Baudrillard:

> It is in the moment of disorder, of loss of differences, when the subject is most miserable, that he perceives most clearly his radical defficiency. He cannot bear this incompletely realized truth and then asks some service from another; it is through the response to that request that the subject's submission and subjection filter in. This is why the instruments of normalization move as closely as possible toward disorder (and here the term is taken in its broadest meaning, encompassing both actual and symbolic instances) to feed on it and produce their specific forms of order. Hence the general principle: the social system produces order from disorder, and most often there is between the new order and the previous state, a surplus of norms analogous to what surplus-value is to capital.[51]

The same balancing of excessive desires and defense mechanisms with which we are familiar in our own lives can be observed also at the societal level, hence Guillaume's paradigm of the "neurotic organization of the social field." But we are dealing here with more than simply a mirror image of individual neuroses insofar as Guillaume maintains that "the societal stock of norms . . . constitutes the reservoir from which everyone draws to organize his own neurotic defenses."[52] In other words, infrastructural crises and their consequences do not account completely for what Guillaume sees as a trend toward the accumulation of power. At a more profound level, individual members of a society express their neuroses in building up individual as well as collective forms of defenses. Guillaume's thesis is that capitalism disrupts the symbolic order more radically than any other historical development. Space is lacking here to restate in detail his analysis of the latest "mode of normalization," which no longer rests primarily on capital accumulation, but involves the manipulation of information in an attempt to extend the field of private appropriation beyond the limits encountered by the industrial system. But his is a pessimistic vision of history. He does not claim that the overthrow of capitalism, or of any power structure for that matter, will allow man to realize his appetitive urges—or, as the Classics used to put it, control his passions.

The weakness of both Attali and Guillaume's theses, and probably of most psychoanalytic approaches to economics or politics, is that they appear to reduce subjectivity to irrationality. There are crisis tendencies in the cybernetic regulation of the production and distribution of informed energy, and at the level of the symbolic values which impose on that process. But there are crisis tendencies of yet another kind which arise from our attempt to make sense of those experiences. Reason may not be as transparent as the Enlightenment philosophers portrayed it, and I will not try to make a valuation of the idea of reason. In particular, the ordering of these different crisis tendencies, which my use of the term "level" seems to suggest, is a philosophical problem that cannot be dealt with here. But a third and distinct category must be included in our classification of societal crises, namely "legitimation crises," to use Habermas' terms, or self-consciousness in my own terms.

For Habermas, social systems can be envisaged from two different points of view. The first paradigm resolves around the process of systems integration, which allows social systems "to maintain their boundaries and their continued existence by mastering the complexity of an inconsistent environment." The alternative paradigm directs our attention to the "systems of institutions in which speaking and acting subjects are socially related";[53] in this context, Habermas speaks of "social integration." He goes on to suggest that these two paradigms can be inter-connected, a view that I obviously share, but he is more concerned with crises of social integration occurring at the level of socio-cultural norms. In raising the problem of the relation of these norms to truth, Habermas brings to the discussion of the notion of order a dimension which was overlooked in the texts reviewed so far.[54]

Norms are inherent in the process of regulation and, in this sense, systems theory is fundamentally normative. It distinguishes among norms, explicates their effects and, above all, permits their manipulation. Whereas the traditional causal paradigm only allows for the manipulation of initial conditions, but not of the "laws" themselves (which is how I understand the meaning of Bacon's aphorism), the systemic paradigm relativizes the intransigency of scientific laws. But on what grounds? This question is of paramount importance in the case of societal regulatory processes, and must be answered before any alteration of the regularities of social behavior is attempted or even advocated.

Socio-cultural norms legitimate coercive structures as long as they are widely shared, thus transforming the use of power into the acceptance of authority, according to M. Weber's classical thesis. In time of crisis, legitimating beliefs ceases to command unconditional adhesion. The analysis put forward by Habermas of the circumstances in which such threats develop is not directly relevant to our topic. But his contention that the selection of new norms can (and should) come out of the exercise of a resonable discourse forces us to consider the possible relations between disorder and reason (subsequent to having dealt briefly with disorder and madness). In particular, we must ask whether the discursive activities in which political actors engage for the purpose of reaching a consensus on the validity of the organizing principles embodied in legal structures, bureaucratic managerial practices, etc., is itself destructive of these very structures. Although raising this question already prejudices the answer (reason connoting critique), it cannot be settled without further empirical research from a variety of perspectives (historical, semantic, etc.). Habermas does not pose this question in this way himself, since he is not concerned with the logic of self-organizing systems. And yet his "advocacy model" of the critique theory of society can be interpreted as a positive, if indirect and hypothetical, answer to it.

Habermas rejects the positivistic separation of facts and values and its corollary that

> practical sentences belong to an autonomous domain that is subject to a logic different from that governing theoretical-empirical sentences and [are] connected with belief acts or decisions, rather than with experiences.[55]

And asserts that the "appropriate model" for determining the validity claims of norms

> is . . . the communication community of those affected who, as participants in a practical discourse, test the validity claims of norms and, to the extent that they accept them with reasons, arrive at the conviction that in the given circumstances the proposed norms are "right."[56]

But the actualization of this communication theory is hindered in advanced capitalist societies by what Habermas calls the "suppression of generalizable interests," meaning those interests that can be expressed as norms based on a rational consensus. Critical theory takes up the task of

> ascertaining generalizable, though nevertheless suppressed, interests in a representatively simulated discourse between groups that are differentiated . . . from one another by articulated, or at least virtual, opposition of interests.[57]

Whether this simulation can be effectively carried out depends on the availability of empirical criteria needed to distinguish among group interests. Habermas acknowledges this difficulty without resolving it, but assumes that several methods can be profitably used for that purpose. The alternative, he argues, would be to accept Luhmann's view that the "old European" orientation toward normative claims is obsolete, and to submit ourselves willingly to the authority of a benevolent technocracy. Avowedly partial to reason, Habermas insists that such an alternative is not inevitable.

Toward a "Crisology"[58]

The immensely rich notion of the unity of opposites has a long genealogy in the history of thought. That it can be applied to social evolution does not have to be proven, unless one could establish, not only that the contributions of Hegel and Marx to social theory were wholly gratuitous, but also that contemporary social science is altogether immune to dialectical reasoning. What needs to be done, however, is to demonstrate that the rediscovery of this notion in the natural science and the "sciences of the artificial," from which it seems to have been completely eliminated, opens the way for a more satisfactory reformulation of the specific traits of social order and disorder. As I have shown, one of the most promising approaches to this problem revolves around the concept of crisis, thematized in the context of a communication paradigm, and scrutinized from the triple standpoint of variations in the semantic codes, imaginary representations, and rational justifications sustaining actual patterns of communication.

Of course, one can look at crisis phenomena from many other angles. The theory of crises, the "crisology" that E. Morin calls for, remains to be attempted.[59] In fact, a theory of crises would certainly open new and interesting perspectives on the problem of alienation. The conditions of existence in complex societies and in the "global village" we are supposed to inhabit, engender an almost unbearable feeling of powerless. Yet complexity is conducive to the emergence of crises, reintroducing chance into our lives and thus the opportunity for free individual choices. True, these fleeting moments of freedom can be canalized and exploited by "The System." We recognize here a favorite theme of the New Left to which Attali, Guillaume, and Habermas all subscribe, to the extent that they are bound by their Marxist view of history on proving that capitalist societies move toward more complete, albeit more subtle, forms of domination. But is it necessarily so? At this point it seems to me that the reference to Marx ceases to be illuminating. Rather than taking up this problem in these terms, once again risking the overstepping of the limits of my subject, I prefer to move the debate to another plane. What is fundamentally at stake here is the logic of evolutionary processes.

> A crisis is not necessarily evolutionary; it can be resorbed by a return to the *status quo*. But a crisis is potentially evolutionary. It bears in itself the characteristics of evolution in a nascent form. To understand this, it is imperative to cast away once and for all the idea that evolution can be linked to a continuous flow. Any kind of evolution always proceeds from events/accidents (perturbations) that give rise to a discrepancy, then to a tendency . . . which entails a series of . . . disorganizations/reorganizations . . . A crisis is therefore a microcosm of evolution.[60]

Now we can apply to the elucidation of the overwhelmingly complex notion of evolution what we have learned from the study of microcosm. *Vice versa*, this macroscopic vision should enable us to achieve a more complete understanding of those aspects that are only nascent in crisis situations.

Notes

1. E. Morin, "Complexity," *International Social Science Journal*, Vol. XXVI (1974), No. 4, p. 559.
2. Ibid., p. 569.
3. See H.A. Simon, *The Sciences of the Artificial* (Cambridge, Mass.: MIT Press, 1969); P.A. Weiss (ed.), *Hierarchically Organized Systems in Theory and Practice* (New York: Hafner, 1971); and A. Koestler, *The Ghost in the Machine* (New York: Macmillan, 1967), among other sources.
4. L. Loefgren, "Structure," in *The Cybernetics of Cybernetics* (Urbana, Ill.: University of Illinois, Biological Computer Laboratory, 1974), p. 104.
5. H.R. Maturana and F.J. Varela, "Autopoiesis: The Organization of the Living," in H.R. Maturana and F.J. Varela, *Autopoiesis and Cognition* (Boston: D. Reidel, 1980), p. 77.
6. In linguistics, the notion of structure can be traced back to F. de Saussure, and it takes a central place in the works of contemporary linguists (e.g., N. Chomsky); in anthropology, one immediately thinks of C. Levy-Strauss; in sociology, structuralism is generally known as structural-functionalism (a nuance which is not without importance), as it has been developed by T. Parsons.
7. F. Varela, "Organization," in *The Cybernetics of Cybernetics*, p. 104.
8. *La méthode* (Paris: Seuil, 1977), p. 104 (my translation).
9. Ibid., p. 105.
10. *A System Analysis of Political Life* (New York: John Wiley, 1965), p. 190.
11. See D.R. Hofstadter, *Gödel, Escher, Bach: An Eternal Golden Braid* (New York: Vintage Books, 1980), pp. 131–134.
12. H. Maturana and F. Varela, *Autopoiesis*, pp. 78–79.
13. This definition is drawn from M. Zeleny and N. Pierre, "Simulation of Self-Renewing Systems," in E. Jantsch and C.H. Waddington (eds.), *Evolution and Consciousness* (Reading, Mass.: Addison-Wesley, 1976), pp. 150–165.
14. Some cellular components, when broken, can be regenerated (e.g., collagen fibers and bacterial flagella); see J.G. Miller, *Living Systems* (New York: McGraw-Hill, 1978), p. 293.

15. See E. Morin, *La méthode*, pp. 133–134.

16. See F. Varela, "Describing the Logic of the Living: The Adequacy and Limitations of the Idea of Autopoiesis," in M. Zeleny (ed.), *Autopoiesis: A Theory of Living Organization* (New York: North Holland, 1981), p. 38.

17. In his preface to Maturana and Varela's paper, Stafford Beer writes that social institutions have an identity "in the biological sense," *Autopoiesis and Cognition*, p. 71; see also E. Jantsch, "The Unifying Paradigm Behind Autopoiesis, Dissipative Structure, Hyper- and Ultracycles," in M. Zeleny (ed.), *Autopoiesis, Dissipative Structures and Spontaneous Social Orders* (Boulder, Co.: Westview Press, 1980), pp. 18–88, and the papers by E. Jantsch, and by G. Pask in M. Zeleny (ed.), *Autopoiesis* (1981), and A. Rapoport, *General System Theory: Essential Concepts and Applications* (Tunbridge Wells, Kent: Abacus Press, 1986), pp. 116–118.

18. See R.W. Ashby, "Principles of Self-Organization," in W. Buckley (ed.), *Modern Systems Research for the Behavioral Scientist* (Chicago: Aldine, 1968), pp. 108–118.

19. "Indeed, wherever one looks, whether outwardly at nature, or inwardly at the thoughts and feelings that are expressions of the operation of the mind, one finds that the essence of things is always in one kind of order or another. Thus, order may well be the basic factor which unites mind and matter, living and no-living things, etc.," D. Bohm, "Some Remarks on the Notion of Order," in C.H. Waddington (ed.), *Towards a Theoretical Biology* (Edinburgh: Edinburgh University Press, 1969), p. 18.

20. M. Maruyama, "The Second Cybernetics: Deviation-Amplifying Mutual Causal Processes," in W. Buckley (ed.), *Modern Systems Research*, p. 306; W. Buckley has also stressed the strategic importance of positive feedback in a system's search for new goals, see his "Society as a Complex Adaptive System," in ibid., pp. 490–513.

21. See E. Schrödinger, *What is Life?* (New York: Macmillan, 1947), *in* passim; for a summary, see H. von Foerster, "On Self-Organizing Systems and Their Environment," in M. Yovits, et al., (eds.), *Self-Organizing Systems* (London: Pergamon Press, 1960), pp. 42–43.

22. Ibid., p. 43.

23. Ibid., p. 38.

24. See S. Coleman, *The Measurement and Analysis of Political Systems* (New York: John Wiley, 1975).

25. H. Atlan, "Du bruit comme principe d'organisation," *Communications*, Vol. 18 (1972), p. 27 (my translation).

26. Considering that Atlan acknowledges von Foerster's influence on the development of his own theory, it is a little surprising that he does not comment on this discrepancy.

27. More specifically: "The *destruction* of order always prevails in the neighborhood of thermodynamic equilibrium. In contrast, *creation* of order may occur far from equilibrium and with specific nonlinear kinetic laws, beyond the domain of stability of the states that have usual thermodynamic behavior," Ilya Prigogine et al., "Thermodynamics of Evolution," *Physics Today*, Vol. 20 (1972), No. 11, p. 24 (the author's emphasis).

28. Erich Jantsch, *Design for Evolution* (New York: Braziller, 1975), p. 37.

29. Prigogine advances that the two complementary aspects one always finds interacting in dissipative structures "correspond essentially to the dialectic of mass and minority" in a social context; "Order Through Fluctuations: Self-Organization in Social Systems," in E. Jantsch and C.H. Waddington (eds.), *Evolution*, p. 126.

30. I. Prigogine and I. Stengers, *Order out of Chaos: Man's New Dialogue With Nature* (New York: Bantam Books, 1984), p. 313.

31. See J. Mehler, "Connaître par désapprentissage," in E. Morin and M. Piatelli-Palmarini (eds.), *L'Unité de l'homme* (Paris: Seuil, 1974), pp. 287–299, and T.G.R. Bower, "Repetitive Processes in Child Development," *Scientific American*, Nov. 1976. On the other hand, H. Atlan has shown that learning causes an increase in the variety of the cerebral system; learning itself results from the interplay of conscious memory and unconscious desires. Furthermore, dreams help to recharge the redundancy potential of the brain. See H. Atlan, "Conscience et désirs dans les systèmes auto-organisateurs," in E. Morin and M. Piatelli-Palmarini (eds.), *op. cit.*, pp. 449–465.

32. See P. Corning, *The Synergism Hypothesis: A Theory of Progressive Evolution* (New York: McGraw-Hill, 1983), pp. 74–75.

33. Ibid., p. 75.

34. So many historical examples seem to bear out this hypothesis that it would be futile to attempt to list them all. But K. Peters offers us an interesting illustration: "Hutterites have institutionalized the creation and maintenance of tensions within their dynamic configurations. Tensions could be reduced in many ways; the reduction of the birth rate, the loss of members, or the lowering of the financial provisions for future generations would be among the most convenient ones. Yet the sect actually strives under pressure that no other population in the world is able to bear." K. Peter, "The Dynamics of Open Social Systems," in J.E. Gallagher and R.D. Lamber, (eds.), *Social Processes and Institutions: The Canadian Case* (Toronto: Holt, Rinehart and Winston, 1971), p. 171.

35. Translators' Introduction to Niklas Luhmann, *The Differentiation of Society*, translated by S. Holmes and C. Larmore (New York: Columbia University Press, 1982), pp. xv–xvi.

36. The transition from an agricultural to a proto-industrial economy in England at the beginning of the 17th century is a case in point. It came partly out of the consequences of an "energy crisis" provoked by the depletion of the forests on the British Isles throughout the 16th century. Far from destroying the British economy, the ensuing tensions made the conversion to a coal-burning industry possible (see J.U. Nef, "An Early Crisis and its Consequences," *Scientific American*, Vol. 237, No. 5, pp. 140–151). From then on, England was propelled into a new era; yet it does not seem legitimate to see in the industrial revolution, which lasted over two centuries, the Nemesis of all British institutions.

37. A. Rapoport, *General Systems Theory*, p. 80.

38. See ibid., Chapter 3.

39. See Gordon Pask, "The Meaning of Cybernetics in the Behavioral Sciences," in J. Rose (ed.), *Progress of Cybernetics I* (New York: Gordon and Breach, 1969), pp. 15–44.

40. J. Habermas, *Legitimation Crisis* (London: Heinemann, 1976), pp. 3–4.

41. See H. von Foerster, "Cybernetics of Cybernetics," *BCL Report*, No. 13-2 (Urbana, Ill.: Biological Computer Laboratory, University of Illinois, May 15, 1975), p. 6.

42. This is certainly true of everyday languages because they are self-transcendent in the sense that they can express meta-statements about themselves without negating their own constitutive rules, which is not the case of formal languages—viz. the famous paradox of the barber who shaves those who do not shave themselves. (On this and other relevant aspects of the use of language in the formation of consciousness, see W. Pankov, "Openness as Self-Transcendence," in E. Jantsch and C.H. Waddington (eds.) *Evolution*, pp. 17–36.) Hence, the cybernetics of cybernetics can only be developed by cyberneticians-cum-philosophers!

43. We are reminded, however, by Crozier and Friedberg that "There are no completely regulated or controlled social systems. The individual or collective actors they consist of can never be reduced to abstract and desincarnated functions. These actors, within the severe constraints imposed on them by the "system," have at their disposal a margin of freedom that they can use strategically in their relations with other people." M. Crozier and E. Friedberg, *L'acteur et le système* (Paris: Seuil, 1977), p. 25 (my translation).

44. For E. Jantsch, the limitation of the concept of information for biologists and social scientists comes from the fact that "information, or phylogenetic history frozen into syntactically ordered form (such as a genetic code, a social law, or a work of art) is "brought to life," or placed in a time perspective, by *instruction* from a higher level of coordination (bringing into play semantics, or the functional relations of a space-time structure to its environment); the result is *intuition*, or learning from within by *re-legio* (connecting through a space-time continuum to one's own origin)—ontogeny recapitulates phylogeny in the life of art or ideas no less than in biology." E. Jantsch, "Evolution: Self-Realization through Self-Transcendence," in E. Jantsch and C.H. Waddington (eds.), *Evolution*, pp. 42–43.

45. *Legitimation Crisis*, p. 1.

46. Much of what follows is drawn from Pierre Gaudibert, "Crise(s) et dialectique," in *Communications*, vol. 25 (1976), pp. 118–127.

47. K. Marx, *Capital*, vol. III (Moscow: Progress Publisher, 1959), p. 249; quoted by P. Gaudibert, *op. cit.*, p. 121.

48. For P. Gaudibert, "[in the case of overproduction crises] forces of destruction as well as of creation [are canalized by] second order negative feedback loops . . . For the capitalist mode of production does not operate by simple adjustments of self-regulating mechanisms: it develops historically according to an evolutionary dynamics. It thus approximates self-organizing systems where noise, disorder, . . . crises become instruments of evolutionary regeneration . . . On the contrary [in the case of the overall crisis of capitalism] destructive forces are less and less contained by second order negative feedbacks. The effervescent irruption of positive feedbacks and the amplification of deviating tendencies . . . are not ineluctably victorious (*op. cit.*, p. 126)."

49. See *supra*.

50. See J. Attali, "l'ordre par le bruit," *Communications*, vol. 25 (1976), p. 94; note that in this text Attali does not include Shanonian messages in his classification.

51. M. Gillaume, *Eloge du désordre* (Paris: Gallimard, 1978), pp. 8–9 (my translation).

52. Ibid., p. 10.

53. See J. Habermas, *Legitimation Crisis*, p. 4.

54. Actually, it cannot find a place in Guillaume's conception of norms which precludes the possibility of their discursive redeemability. Indeed, he ends his essay with a plea for the overcoming of all finite meanings ("*le dire*") by praxis ("*le faire*").

55. *Legitimation Crisis*, p. 102.

56. Ibid., p. 105. Habermas's position is not without parallels. Interestingly, in view of this author's sweeping condemnation of the epistemological status of norms in systems theory, a renown systems theorist has given to the problem of the normative dimension of planning a solution which echoes that of Habermas himself. H. Ozbekhan concurs with J. Bronowski in stating that "verification is fundamentally a social process rather than a feeling forever sealed within a person, for the reason that 'truth' is a social fact" (p. 76) and adds that "value propositions have empirical meaning, namely . . . they can be justified by evidence which is socially held to be true" (p. 77); consequently Ozbekhan defines normative planning as a social process whereby both the environment and the value system are made "consonant" (p. 153). H. Ozbekhan, "Toward a General Theory of Planning," in E. Jantsch (ed.), *Perspectives of Planning* (Paris: OECD, 1968), pp. 46–155.

57. Ibid., p. 117.

58. I borrow the term "crisology" from E. Morin; see his "Pour une crisologie," *Communications*, vol. 25 (1976), pp. 149–163.

59. See ibid. However, it would be good to keep Luhmann's admonition in mind (see f.n. 35).

60. Ibid., p. 162.

4

From Biological to Societal Evolution

Evolutionary interpretations of the human condition are certainly not uncommon. Sociobiology, and the debates that it has provoked, is only the most recent example of this genre. We have witnessed also the emergence of a new subfield in political science, somewhat inappropriately called "biopolitics." Nevertheless, there are still very few examples of detailed epistemological critiques of the relevance of evolutionary biology, not to mention Prigogine's evolutionary physics, to political inquiry—Peter Corning's seminal work *The Synergism Hypothesis*[1] being a notable exception. Ever since Darwin's time, theories of biological evolution have given rise to so many instances of misunderstanding and misuse (e.g., Social Darwinism) that a detour through the life sciences is required in order to clear the way for a discussion of societal evolution. While I do not disagree with Corning's thesis that "Darwinian teleology" together with the paradigmatic notion of synergism—i.e., the combinatorial or cooperative effects that facilitate survival and reproduction—can explain evolutionary change, including political development, my strategy here is somewhat different. I argue that biological evolution, systemic differentiation and growth are isomorphic processes answering to an overarching logic, namely the ontogeny (i.e., individual development) of autopoietic unities. Evolution is the history of the sequential reproduction of living systems:

> what evolves is pattern of autopoietic realization embodied in many particular variations . . . thus, the individuals, though transitory, are essential, not dispensable, because they constitute a necessary condition for the existence of the historical network which they define. The species is only an abstract entity in the present, and although it represents a historical phenomenon, it does not constitute a generative factor in the phenomenology of evolution, it is its result.[2]

Some biologists may feel uncomfortable with this position, but I consider it to be sound from an epistemological standpoint. The first section conveys the concerns of those authors who think that evolution is not simply a matter of "chance and necessity," as Jacques Monod said, but is influenced also by the active development of autonomous living beings. While these ideas may not be consistent with the most dogmatic interpretations of contemporary genetic theories, they are not meant to be a reappropriation, and even less a vindication of Lamarckism.

The significance of this paradigm is that it opens onto the socio-cultural level and thus enables us to recover the once discredited concept of societal evolution. The second section illustrates how the latter allows us to pose the problem of social change in terms of the actuation of human potentials, often mediated by political communications, and critically re-interpreted by social theory.

The Dialectic of Order and Disorder in Nature: The Production and Reproduction of Life

Our previous discussion of the creative function of disorder points toward an underlying logic of historical change. I use the term "logic" to underline the abstractness of such an approach. A process remains an abstraction as long as it cannot be shown to manifest itself, to borrow a phrase from A.N. Whitehead, in "occasions of experience" for the observer. Only structures, as defined earlier, have a concrete existence and can be located in space and time. "Experiences" are occasioned by the interaction of structures; immanent dynamic processes are either imaginatively created informing our perception of concrete realities or inferred a posteri. This does not mean that science should only be concerned with "things," a view severely criticized by C.H. Waddington,[3] but rather that processes always need to be contextualized; or, as Whitehead has put it, even though an occasion of experience is related through space and time to everything else, its specificity depends on a limited set of constituent factors, which he called its "concrescence."[4]

This is precisely what Marx tried to do in applying the dialectical method to his analysis of man's productive activities. He failed, however, to investigate this profound intuition far enough. Without denying the specificity of human societies and the uniqueness of man's historical consciousness, it is both possible and imperative to regard societal "infrastructures" and "superstructures" as particular examples of the general class of autopoietic machines. Following Morin, "machines" are entities whose productive activities are not limited to processing inputs drawn from their environment, as is the case with the human artifacts we call machines; machines may also be self-reproductive. We are surrounded by such machines, although we often fail to perceive them in their spatial and temporal unity, and reduce them to the status of passive objects—the sun and all the stars, the ecosystem, etc.[5] Thus such a term "machine" does not represent a return to the mechanistic approach in which the processing of things is analogous to the production of effects by antecedent and external causes. An autopoietic machine produces its own causality, not in complete isolation from, but in relation with, other machines.

The usefulness of the theory of self-organization by simultaneous destruction/restructuration must be judged by the extent to which the theory furthers our comprehension of the performance of the actual physical, biological and societal machines producing the conditions of our concrete existence. Before testing this instrument on the societal "mega-machine" (Morin), it seems prudent to try our hands, so to speak, on a perhaps not simpler, but at least slower-paced machine: the reproduction of life. A more substantial reason for this digression is that all socio-cultural systems have been generated by biological systems. Again, the fundamental difference between the rudimentary man-made allopoietic machines and natural autopoietic machines is that the latter possess an "organizational generativity" which the former lack.[6]

The Archetypical Machine

Life is an archetypical, if not *the* archetypical, machine. Indeed, according to François Jacob, the realization, in the second half of the 18th century, that animals are "organized bodies" marks the beginning of modern biology. By contrast, Descartes' definition of living beings as (allopoietic) mechanisms could only lead to serious misconceptions about the phenomenon of reproduction.

> As long as living organisms were perceived as combinations of visible structures, preformation provided the simplest explanation for the persistence of these structures through succeeding generations. . . . In the second half of the eighteenth century and the beginning of the nineteenth, the very nature of empirical knowledge was gradually transformed. . . . It was within living bodies themselves that the very cause of their existence had to be found. . . . Form, attributes, and behavior all became expressions of organization. . . . Thus . . . a new science was to appear. Its aim was no longer to classify organisms, but to study the process of life; its object of investigation was no longer visible structure, but organization.[7]

To this day, organization remains the cornerstone of biology. According to E. Sinnott, "[an understanding of organization is] the biological problem to which every other is subordinate and contributory."[8] The paradox is that, while the recent and spectacular progress of molecular genetics—and the reductionist tendencies associated with such developments—distract biologists from the phenomenology of living organisms, they also provide them with the opportunity to make again the concepts of reproduction, and organization paramount. We still lack, however, a fully articulated and coherent theory to take advantage of this opportunity. Few authors have gone as far as Maturana and Varela to claim that autopoiesis is a necessary and sufficient condition for the occurrence of all biological phenomena,[9] which I consider to be a dogmatic overstatement. Nevertheless, it is a suggestive hypothesis in the light of which I propose to evaluate those other approaches that more or less approximate it—or depart from it.

The Evolution of the Theory of Evolution

Biological evolution involves a complex interplay between the dynamics of the genosphere and of the phenosphere. The former combines the replication, recombination and reconstitution of the genetic material; the latter works through natural selection. The crucial linkage between the genosphere and the phenosphere is realized in the process of epigenesis.[10] Historically, these aspects have not always been properly balanced in theories of biological evolution.

The maturation of geology and paleontology towards the end of the 18th century laid the foundations for the evolutionary theories of Lamarck, G. Saint-Hillaire, and later of Wallace and Darwin which all emphasized, in their respective ways, the transformation occurring in the phenosphere. Thus, for Darwin, unequal survival and reproduction rates due to functional differences in the capacity to utilize available resources cause, over time, gradual changes of populations as well as the production of new species. Too little was known then about genetic factors. Darwin never heard of Mendel's discovery (in 1865) that the transmission of hereditary information is made by discrete and replicable units. This was later rediscovered by H. deVries, among others. From these beginnings, W. Bateson, T. Morgan, and their followers (especially in Great Britain) developed the new science of genetics which, for a while, relegated natural selection to an almost prescientific status. The experimental proof that genetic mutations occur independently of any identifiable environmental pressure was posed as a dogma. The transformations of the genotype of individual organisms became paramount. The pendulum had clearly swung to the other extreme.

During the 1930s and 1940s, T. Dobzhansky, T. Huxley, and E. Mayr, among several other evolutionists, worked out the so-called neo-Darwinist synthesis.[11] It shifted the direction of genetic research from an exclusive preoccupation with individual genotypes to an analysis of the dynamics of the gene pools of large populations upon which natural selection exerts its influence. Watson and Crick's subsequent discovery of the structure of

DNA, although crucial, prevented the development of the new synthesis into a truly holistic perspective. This discovery largely accounts for the methodological reductionism in contemporary biology, as exemplified by J. Monod's *Chance and Necessity*.[12] However, the parallel advance of population dynamics and ethology has barred a return to the strictly genotypic paradigm of the 1920s. Still, too little efforts are devoted to the comprehension of the epigenetic process. This has led to a misdirected cybernetization of evolutionary biology. It is not difficult to see through it an equally misdirected cybernetication of societal evolution.

At first sight, the neo-Darwinist synthesis seems to "fit" the von Foerster-Atlan model remarkably well. The nucleic acids, DNA and RNA, consist of chains of nucleotides consisting of two invariant and one variable elements (nitrogeneous compounds) which often are called the "letters" of the genetic message, itself expressed in sequences of three-letter "words" (three nucleotides) called "codons." Despite the high redundancy of the genetic code, some noise can interfere with the transmission of the genetic message. It is due to the occasional misspelling or inversion of words, or even to the disruption of phrases (at the chromosomic level). Noise is a very appropriate term, since randomness constitutes the only possible agency of change. This genetic variability is then filtered out by natural selection which directs the course of phyletic evolution and speciation.[13] Natural selection is a complex process involving both conflictual and cooperative interactions; even if the context of Darwin's own work, the "struggle for existence" is a loose metaphor, and ethology has since vindicated Kropotkin's views about the selective value of cooperative behaviors.

The importance that the orthodox theory of evolution attaches to chance has, of course, given rise to critical reactions. Some critiques originate in a misunderstanding of the neo-Darwinist theory which is too often reduced to its Mendelian aspects. J. Monod took pains to explain that the theory recognizes chance and "necessity," i.e., natural selection, as being equally significant controlling forces.[14] It should be added though that some biologists doubt that the synchronic random variations of the genetic material revealed by experimental observations and the diachronic mutations recorded by paleontology are strictly analogous phenomena.[15] But this is a technical issue that can only be settled by biologists and upon which I cannot intelligently comment. A priori, there is nothing exceptional or inacceptable about the fact that highly complex systems like animal populations operate under conditions of uncertainty; indeed the contrary would be very surprising. The contentious issue concerns the concrete location of the interface between genetic order/disorder and ecological order/disorder.

According to the new synthesis, the self-organizing properties of life are best apprehended at the level of the gene pool of a population. But is it valid to consider the gene pool as a fully autonomous autopoietic machine?

This is a crucial epistemological, rather than technical, issue. Its domain of validation cannot be left entirely to the biologists. The relevance of the new descipline known as "sociobiology" hinges on the properties that theoretical biology rests in the gene pool of man; conversely, an inquiry into the sociobiological implications of the attributes conferred on the gene pool could and should inform the way in which these very attributes are defined. That genetic determinations strongly influence animal or even human behavior and social interactions is an obvious, albeit sometimes belittled, fact. But the problem is to know how these determinations are evolved. The neo-Darwinists argue that the physiochemical variation, on the one hand, and the selective pressures of the environment on the other, operate without mediation on the gene pool of a population. This is precisely what the epigenetic or "neo-organicist" paradigm refutes. In effect, where Monod could only perceive chance and necessity, we have also to look for freedom.[16]

Beyond Chance and Necessity

> The mere hereditary transmission of mutable genetic information is not really good enough to qualify as life. We need something that can really undergo evolution by the process of natural selection, and that involves something which interacts with the environment sufficiently for natural selection to be effective. . . . Life which is to engage in evolution requires in fact not only a genotype but a phenotype. The process by which the genotype becomes worked out into the phenotype can be referred to as "epigenesis."[17]

C.H. Waddington did more than any other biologist to further our understanding of this process, but it would be quite wrong to regard the epigenetic paradigm as Waddington's idiosyncratic vagary. Besides, J. Piaget has extended its application to the study of cognitive development and more generally to espistemology. I hereafter recapitulate Waddington's working hypotheses, then indicate a few other convergent analyses before attempting a synthesis centered on the integration of the dual notions of invariance and change which is emerging in the conceptual system of modern biology.

The genetic instructions which an organism inherits are not a kind of Shanonian message with a fixed amount of information, but an algorithm whose completion through time produces a phenotype containing *more* information than its genes. Waddington used a mathematical analogy to illustrate this point; the genetic message resembles a set of axioms, the resulting phenotype resembles a fully articulated system such as Euclidian geometry.[18] J. Monod considered this to be a "false paradox." He explained that the genetic message is equivocal in the sense that it can be "interpreted" into a large variety of globular structures, but the initial conditions of its transmission eliminate all but one of these alternatives. Thus, the genotype determines the phenotype even if it does not specify it a priori.[19] From a methodological standpoint, this formulation is unworkable because it singles out the genetic factor as being the only significant one, the remainder being cast together into a residual category. The primary problem is to integrate both the genetic message and the condition in which it is transmitted into a coherent model that could, in particular, account for simultaneous changes in the initial conditions. This is precisely what Waddington attempted to do. The elementary process of epigenesis, namely cellular differentiation and changes in form (morphogenesis) are canalized by "chreods"—a concept introduced in the previous chapter; this means that they are not oscillating around an equilibrium point, but change in a controlled manner. Both the genotype and the environmental pressures determine the changing shape of the developmental pathway. Therefore, the interactions between the phenotype and the environment do not take place on a random basis, as the neo-Darwinists would have us believe. In C.H. Waddington's words:

> The environmental factors which exert natural selective pressure do not . . . have any important tendency to produce appropriate changes in the genotype, but they do very often tend to produce appropriate alterations of the phenotype, and it is on the phenotype, not on the genotype, that natural selection operates.[20]

From this standpoint, the problem of the inheritability of acquired characteristics returns to the forefront of evolutionary theory. Of course, Waddington did not propose to revamp the Lamarckian paradigm. No matter how hard a giraffe tries to extend its neck, its offspring will not be born with longer necks. Natural selection works on populations, not individuals. In a large group confronted with exceptional environmental circumstances, some phenotypes will adapt to these circumstances to the extent allowed by the kind of genetic variability available to the group. The point is not that the environment directly selects the hitherto unexploited genes but rather it offers a challenge to which the organisms respond behaviorally and/or morphologically.[21] Organisms are presented a series of "choices"—escaping predators by running faster or migrating where predators are less numerous, for example—and each alternative constitutes a different filter for the genetic material contained in the gene pool (itself subject to random variations). Because the epigenetic processes are resistant to change,

> The development of the adaptive phenotype may itself [over a long period of time] become buffered and resistant to latter alteration when the original precipitating circumstances are removed. . . . Thus a combination of the

> epigenetic paradigm with a general population outlook leads one to expect the occurrence of a process which, on the population level, gives a result similar to what the inheritance of acquired characteristics in the classical sense was supposed to give on the individual level.[22]

In brief, the phenotypes produce, at least in part, the conditions of their own selection. The circularity of this argument lends itself ideally to a cybernetic analysis.[23]

> We would have to contemplate a subtle system of feedbacks, in which the environmental factors which exert natural selection operates on phenotypes which they themselves have influenced in relevant ways (usually to improve their effectiveness), and in which the phenotypes in their turn have an influence on the nature of the natural selective pressures to which they will be subject.[24]

L.L. Whyte's notion of internal selection is quite compatible with Waddington's epigenetic paradigm.[25] Internal selection realizes the "satisfaction of coordinative conditions" within individual organisms and determines the type of "evolutionary material" that environmental pressures can select.

The hypothesis of the hierarchical structuration of the biosphere is implicit in the epigenetic paradigm. It is rendered explicit by Arthur Koestler who was quite familiar with both Waddington and Whyte's works. Koestler claimed that complex systems consist of a hierarchy of autonomous "holons," that is to say, "sub-wholes" behaving towards their subordinated parts as self-contained wholes, and towards their superior controls as dependent parts.[26] To the extent that such an approach is derived from L. von Bertalanffy's theory of open systems, it may seen to be better suited to the investigation of embryological processes, or the physiological performance and behavioral patterns of biological organisms, than to the explanation of their evolutionary development. Yet Koestler has suggested that it is relevant to the analysis of both ontogeny and phylogeny.

Koestler strongly rejected the neo-Darwinist theory because it does not pay sufficient attention to the internal coordinative conditions that channel the consequences of random mutations. The mutations "would be submitted to the sifting and processing at the gate of successively higher levels of the hierarchy." He concluded that "there must be unitary laws underlying evolutionary variety, permitting unlimited variations on a limited number of themes."[27]

As far as external evolution is concerned, Koestler emphasised the creative, innovative character of animal behavior: living beings do not passively respond to their environment but alter it to suit their needs. Their "exploratory drive" helps them to acquire new skills, some of which are passed on from one generation to the next. In this respect, Koestler mentions

Waddington's hypothetical model of "genetic assimilation" to which we have already alluded; but he appeared a little too eager to resurrect Lamarckism from its scientific limbo, ignoring Waddington's methodological doubts.[28] As stated earlier, the logical foundations of the theory of evolution can and should be a concern for both biologists and non-biologists. In this respect, Koestler was right when he underlined the contradiction inherent in modern biology which acknowledges the purposiveness of ontogeny but refuses to extend it to phylogeny, even though, as he put it, "phylogeny . . . is a sequence of ontogenies."[29] However, it is not proper for amateur biologists to dictate specific research assignments to professional specialists. A return to Lamarck is probably not the best way to approach the purposiveness of phylogenetic processes, unless we decided to qualify as Lamarckist any theory that casts a doubt about the universality of the Mendelian dogma.[30]

It is remarkable that a more orthodox scientist, François Jacob, (Nobel Laureate, 1965) has felt the need to formulate a concept comparable to that of "holon" to account for the complexity of life:

> Even the simplest organism is so complex that it could probably never have taken shape, reproduced and evolved if the whole had had to be built piece by piece, molecule by molecule, like a mosaic. Instead, organisms are built by a series of integrations . . . [thus living beings] are arranged according to a hierarchy of discontinuous units. . . . Each of these units formed by the integration of sub-units may be given the general name "integron." An integron is formed by assembling integrons of the level below it; it takes part in the construction of the integron of the level above.[31]

What makes complex living systems adaptive? Peter Corning argues that the answer to that question lies in the concept of "functional synergism," i.e., "combinatorial or co-operative effects that have had positive consequences in relation to various aspects of the problem of survival and reproduction."[32] Synergism results in significant qualitative changes. He expresses this in the form of a hypothesis:

> It is the selective advantages that arise from various synergistic effects that constitute the underlying cause of the apparently orthogenetic (or directional) aspects of evolutionary history, that is, the progressive emergence of complex, hierarchically organized systems.[33]

Admittedly, this may look like a tautological formulation. However, Corning provides many examples of actual synergetic effects which show clearly that such phenomena can be empirically observed (e.g., eukaryotic, i.e., structurally complex nucleated cells developed as a consequence of the symbiosis of components like the mitochondria or chloroplasts which had independent origins).[34]

It would be quite erroneous to conclude that the epigenetic paradigm postulates the existence of a pre-established harmony unfolding through time in a linear progression toward perfection, simply because it does not assign a preeminent role to random variation at the molecular level. On the contrary, Waddington proclaims:

> It introduces into the theory [of evolution] an inescapable indeterminism quite different in nature from quantum indeterminacy, but almost as intractable, since identical phenotypes may have different genotypes, and identical genotypes may give rise to different phenotypes.[35]

The complex interactions developing between these two poles provide innumerable opportunities for the "branching out" of evolution. The adaptive competence of the phenotypes can perhaps be best understood in terms of a learning process.[36] P. Corning cites an impressive list of empirical studies in support of a very similar conclusion; for example,

> To make sense of the reproductive cycle of herring gulls . . . an entire arsenal of concepts and analytical tools is necessary, starting with a framework of hierarchically organized goals and subgoals . . . and including memory and information-processing functions, learned behaviors and skills, sensory input and feedback functions, benefit-cost analysis, optimization theory, and probabilistic decision analysis.[37]

In the case of the higher organisms, the lessons of evolution are memorized in, and exploited by, complex central nervous systems that can accelerate the pace of evolutionary change. Thus goal-oriented cultural systems and their organization into a network of autonomous political relations do not evolve outside of natural processes, but in harmony with them. We see, therefore, that at the biological level, no less than at the societal level, "noise" is not a simple notion. But we also see, and more clearly at the former than at the latter level, how and why noise is potentially creative. To reproduce the same, in a world constantly changing under the contradictory influences of entropic decay and autopoietic regeneration, requires the incessant production of differences by autonomous unities whose individual existences are coupled in networks evolving into differentiated group structures identified by a subjective observer.

Invariance and Change in the Reproduction of Life

We have seen that evolution involves more than the transmission of information coded in the DNA. But this does not mean that evolution and reproduction are completely distinct concepts. Obviously, species must reproduce to evolve. Such a tautology can lead to the less trivial research hypothesis that reproduction and evolution are isomorphic processes, operating on different time scales but explainable in similar terms. More specifically, Yves Barel argues that reproduction, growth, and evolution correspond to different ways of conceptualizing the same thing, i.e., the realization of the natural dialectic of invariance and change in the self-organization of life.[38]

Barel's argument is based on an extensive survey of the literature in biology and systems theory. While his demonstration is certainly consistent with the tenets of the epigenetic paradigm, he avoids the biased attacks upon conventional neo-Darwinist theorists in which Koestler, and to a lesser extent Waddington, too often indulge. In fact, if the epigenetic paradigm has any value, it should lead to synthesis of the array of biological knowledge accumulated over the last two decades, rather than to a new, orthogenetic (i.e., finalistic) dogma.

First, growth and reproduction are complementary and partially identical phenomena. Their interrelation is mediated by a third term, namely differentiation. In a certain sense, reproduction and growth are identical phenomena.

> When the cells of an organism reproduce, one says that the organism grows. Similarly, a population increases . . . when its individual members multiply. Growth is then the result of the process of reproduction.[39]

But we know that biological reproduction almost always implies some degree of change; for example, sexual reproduction does not produce strictly identical replicas of either one of the parents. On the other hand, embryology has established that, as their number increases, the cells contained in the egg cease to reproduce identically (re: the process of "gastrulation"), and from then on epigenesis becomes more and more complex. Therefore,

> growth is as much the result of the relative importance of the differentiations generated by reproduction than of the very nature of these differentiations for the performance of systemic functions on another level of integration.[40]

In other words, reproduction and growth are identical phenomena insofar as they imply multiplication and differentiation, but they correspond to separate levels of analysis.

Second, evolution represents an extension of the complexity scale upon which reproductive processes are realized. (In what follows I stay less close to the letter of Barel's writing but I do not think that I distort the spirit of his argument.) The fundamental contribution of neo-Darwinism is that it explicitly posits a correspondence between reproduction and evolution. Its Mendelian component establishes that evolution cannot proceed in all directions at once, nor even in one arbitrarily selected direction as Lysenko tried in vain to prove, but results from the production of reproducible differences in the arrangement of immutable discrete elements. The reproducibility of these differences implies that biological evolution is never a complete revolution but an endless recombination of perduring systems (for instance, we still find in the human brain the remnant of reptilian structures). On the other hand, natural selection operates on the basis of the reproductive capacities of interacting populations. What the neo-Drawinists fail to perceive, however, is that their master concept is not an objective construct but the projection on a given plane of a complex and irreducible system of ideas—what Barel calls the dialectic of invariance and change. We have already seen that, on another plane, we ought to speak of development or growth. Evolution subsumes and transcends reproduction and development, but at such a level of complexity that it is perhaps impossible to sustain a coherent analysis at this exalted level.

Teilhard de Chardin attempted to carry out such a project; we lack validation criteria to determine whether or not he succeeded in his endeavor. This is precisely the reason why I have deliberately avoided discussing the thorny question of where biological (or any other form of) evolution ultimately leads to. But we can suggest with Barel that living organisms evolve through the interaction of their genotypic instructions, of their epigenetically developed phenotypes, and also of the entire time-space.[41] Evolution thus realizes an immensely complex totality that cannot be wholly apprehended at the level of the gene pool (as some neo-Darwinists would have it), nor even at the epigenotypic level (as Waddington claims). Furthermore, this totality manifests a quasi-intentional *démarche*, in the sense that life has succeeded so far in reproducing itself, however improbable the conditions of such a reproduction may often appear to be.

Barel's strategy of conceptualizing reproduction as a means of understanding the dialectic of invariance and change in life and in society, rather than as an unequivocal reality, appears to be more promising than Teilhard's grand scheme. Perhaps the epigenetic paradigm aims at a very similar goal but at the level of development rather than at the level of reproduction. Such a distinction, however, is somehow artificial, all the more so because Barel explicitly adopts many of Piaget's epigenetic assumptions. Given the logical connections linking reproduction, development, adaptation, and evolution, it is arbitrary to define reproduction as the key to this irreducible cycle. The representation of life as an autopoietic machine helps to justify such a choice.

Living organisms produce a large number of proteins and other chemical compounds, they produce artifacts, tools, social habits and even, in the case of man, they produce languages and cultures. Reproduction occurs when these productive operations are brought to bear upon themselves by epigenesis. Epigenesis or development occurs when reproductive processes are brought to bear upon themselves by adaptation and evolution. Evolution occurs when developmental transformations are brought to bear upon themselves by . . . reproduction. In a similar vein, Maturana and Varela claim that "evolution . . . is necessarily secondary to the establishment of reproduction," and "reproduction requires the existence of a unity to be reproduced"; hence "the proper evaluation of the phenomenology of living systems . . . requires their proper evaluation as autopoietic unities."[42]

The problem is: What can we know of the productive and reproductive performances of autopoietic machines? According to Morin, there is simply no way of objectifying these machines. Either we model them on allopoietic machines, as the first cybernetics did, and our perception is limited to their regulatory, functional and pre-programmed performances. Or we model them on Prigogine's dissipative structures, or on what Morin calls the universal "arkhe-machine," and we see in them the manifestations of disorder, or antagonistic and aleatory processes destroying and regenerating unstable systems. Real physical machines, however, oscillate between these two poles. Biological and anthroposocial machines, on the other hand, occupy a multipolar space because their unity is complex, paradoxical. As it is the case with subatomic particles, the observer cannot perceive everything at once but can only consider a particular machine from one standpoint at a time. Therefore,

> He must have at his disposal a method that could allow him to conceive the multiplicity of standpoints, and then to move from one to the other; he must have at his disposal theoretical concepts that allow him to do so fruitfully, instead of isolating physical, biological, and social entities.[43]

In conclusion, our survey of contemporary evolutionary biology suggests that a thorough epistemological analysis of biological reproduction should be the first step towards such a method. To produce etymologically means to lead forward (i.e., *pro-ducere*). Biological (re)production leads forward to some identifiable, and therefore invariant (at least for a while) objective: the perpetuation of the species. Genetic and epigenetic determinations contribute to achieve this imperative. But to the extent that biological (re)production *leads* towards something, it follows a course of its own. The genetic code randomly breaks down at times; species "choose," as Waddington writes, the conditions of their own selection, primates and man engage spontaneously in creative activities. To produce is a conditioned act, but an act nevertheless. To reproduce life is both a necessity and a game—of chance and of skill. In other words, the stability, reproduction and adaptation of autopoietic machines never conform to an objective necessity, i.e., to a transcendent law of evolution.

At this point "the extraordinary similitude of language between biology and the social sciences" becomes an inescapable conclusion; Barel goes on to remark:

> The paradox is that it is the very concept of reproduction which introduces the most important difference in the vocabulary of biology and of the social sciences, owing to the important role it plays in the former, whereas the latter have made a much lesser use of it.[44]

The similitude is obvious in the case of concepts like growth, change, differentiation, assimilation, evolution, etc. It is tempting to posit that

> . . . the biological system is in part but the application to biology and by biologists of a broader system [that is] common to biology, the social sciences, and to still other disciplines.[45]

For Barel, the rudiments of this overarching system are to be found in cybernetics, information theory, and general systems theory. So far I have attempted to show that within these approaches, the emerging theory of self-organization is much more relevant and promising than earlier models; this should become even clearer in the context of social evolution.

Societal Evolution: Myth or Reality?

As it was suggested above, evolutionary biology points toward a number of parallels between living and societal systems. The concepts of autonomy, choice and even freedom are meaningful in both the biosphere and the universe of social relations. However, this apparent semantic convergence between biology—or, at least, those theories that are built upon the epigenetic paradigm—and socio-political theory does not settle the issue at stake here, namely whether or not social evolution (i.e., directed social change) is a fallacy. Until relatively recently, most social theorists hesitated to use the phrase "social evolution" lest they would be found guilty of association with Herbert Spencer's monumental failure to separate science from ideology. Today though there is a renewal of interest in this notion, in part because of fascinating breakthroughs in anthropology and paleontology.[46] Moreover, the new subfield of biopolitics offers challenging opportunities for articulating evolutionary biology with political inquiry; according to S.M. Hines, one of the most important tenets of biopolitical research is that "The study of politics must reflect that human beings are an outcome of evolution."[47]

My concern here is with the problematic validity of an evolutionary perspective in the social sciences. Of course, societal systems are constrained by some of the same ecological and thermodynamic organizational principles at work in the biosphere. The human species evolved as a result of the interaction of many of the same factors that influenced the development of all the other forms of life (e.g., genetic mutations and synergetic physiological and/or behavioral effects). And cultural innovation can be understood as a trial and error process comparable in some ways to other types of learning situations, for example in animal societies. However, the hyper-complexity of the societal sphere, which is due to the unparalled cognitive capacities of the human actors, confronts the student of politics with a unique problem. That is, what can be regarded as meaningful (evolutionary) change? I argue that invariance and change are, to an irreducible extent, arbitrary notions. I suggest also that a necessary, although not sufficient condition for the construction of a minimally arbitrary theory of societal dynamics is to conceptualize change in societal relations with reference to an evolutionary paradigm—i.e., a research strategy aimed at the reconstruction of an "inner logic" in the structuration of societal processes. (By "minimally arbitrary" I mean to describe a situation in which the identity of the interacting actors, as well as the identity of their collective creations, is not denied.) A crucial and thorny problem in this respect is how to define the sociocultural equivalent of biological adaptation.[48] My answer is as follows: i) evolution, insofar as it means the history of the species, is not an objective notion, but a reconstruction and, therefore, adaptation should not serve as a screen behind which social theorists hide their values; ii) biological organisms, as we have seen, adapt to circumstances that are, in part, of their own choosing; iii) the degree of freedom that characterizes human adaptive choices is evidently unsurpassed in nature but, at the same time, the concept of adaptation forces us to relate this freedom to the experience of basic human needs. This provides me with an opportunity to review critically some of the most significant contributions to a theory of societal evolution. My own standpoint is that i) sociobiological models that tend to reduce societal interactions to genetic mechanisms are inadequate, even if the possibility of some kind of reaction of behavioral factors upon the selection process is acknowledged; and ii) inversely, sociological theories that pay too little attention to evolutionary biology tend to emphasize arbitrarily selected aspects of social change, since they lack criteria for distinguishing societal hyper-complexity from the complexity inherent in all living systems.

The Problem of Change

To state that everything changes is logically equivalent to stating that nothing changes.[49] Any theory of the dynamics of natural or social processes must be based on an axiomatic characterization of the categories of invariance and change. This need not take the form of a timeless dichotomy, nor does the axiomatic nature of this partition necessarily entail a decisionistic operation ("values are a matter of personal preference"), since it does not preclude the possibility of a "discursive redemption" (Habermas). What it means, rather, is that we must posit simply that "the environment of the observer [i.e., reality] is what it is,"[50] and that the properties of invariance and change are conditional upon the act of observing and constructing objects of observation and not properties of the objects themselves. Moreover, the objects and the events associated with their observations are themselves created by the properties of the chosen representation. Consequently, the environment is *perceived* as the locus of stationary, moving, or changing objects. But these perceptions occur within an organism; they are manifest in its acts and the logical structure of internal representations emerges from the logical structure of the corresponding acts. Thus, the information contained in a representation of the world is operative only to the extent that it can be related to the cognitive structure of the subject to whom it is addressed, and also on condition that it can be converted into meaningful action. For instance, biological evolution is an informative representation because the preservation of organic life ranks high in the "tacit knowledge" (Polanyi) common to all human beings. It is certainly not specious to consider that biological evolution could be a fallacy. However, it seems perfectly rational to interpret change within the biosphere as meaning the enhancement or degradation of life.[51] By contrast, the evolution of society poses much more serious problems. Qualitative transformations of social interactions and cultural values can be described with reference to a wide range of criteria and many representations of these processes have been advanced. To simplify the discussion, I will classify these into four categories, namely the mechanicist, historicist, functionalist, and genetic-structuralist paradigms. I argue that only the genetic-structuralist paradigm can yield an understanding of evolutionary processes consistent with the epistemological position I have taken with respect to what I hereafter call "the problem of change."

The mechanicist approach combines the Cartesian analytic method with the positivist separation of facts and values. It does not address directly the problem of the meaning for the observer of any variation in the relations posited by theoretical models and validated by empirical observation. In other words, the mechanicist paradigm ignores the representational character of scientific theories, thereby confusing a constructed image of the world with reality—to quote Whitehead again, it rests on the "fallacy of misplaced concreteness." Thus it offers no strategy for assessing critically the arbitrariness inherent in its determination of what constitutes factual evidence for it reduces reason to the scientific method. A criticism of this kind can be applied to the non-evolutionary paradigm of the behavioralist movement of the 1950s and 1960s which was inspired from psychology rather than from biology. (For example, changes in party identification are explained by reference to other behavioral patterns and/or environmental circumstances, themselves shaped by other behavioral patterns and other environmental variables in an infinite regression.) However, the mere adoption of biological concepts and of an evolutionary paradigm is in no way an antidote against mechanistic tendencies, as the case of sociobiology clearly illustrates. The definition of sociobiology provided by one of its best known founders is very unhelpful, for there is more to sociobiology than merely "the systematic study of the biological basis of all social behavior"; much more, indeed, since we are told immediately thereafter that "sociology and the other social sciences . . . are the last branches of biology waiting to be included in the Modern Synthesis."[52] The works of R. Dawkins, R. Trivers, E.O. Wilson, among others, have popularized the notion that in animal as well as in human societies behavioral patterns are the result of genetic determinations consistent with the imperative of reproductive success, just as it is the case with anatomical and physiological traits.[53] (For example, kinship-based "altruism," which involves some form of deprivation or even self-sacrifice in the interest of genetically related individuals, can be found at many levels of organization and in many species; a much more controversial argument is E.O. Wilson's contention that there exist genes for conformity, xenophobia and aggressiveness.)

The sociobiological program has been criticized from at least three vantage points. First, it is based upon a rather dogmatic interpretation of the new synthesis which does not take into account synergism, learning, etc. . . . , that is, the complex interactions between the genotype and the phenotype which were discussed in the preceding section.[54] Second, as M. Sahlins, among others, has pointed out, cultural rules do not correspond to what genetic determinations would lead us to expect to find when studying actual societal dynamics.[55] Finally, the Science for the People group (e.g., S.J. Gould, R. Lewontin) has accused sociobiology of being a reactionary ideological discourse.[56] Wilson, in collaboration with the physicist Charles J. Lumsden, has modified his views to a limited extent in order to deal with the anthropologists' argument that cultural influences are much more directly relevant to the explanation of social behavior than genetic factors. These authors propose a coevolutionary model in which natural selection operates on both ordinary genes and what they call "culturgens," i.e., "relatively homogeneous group[s] of mental constructions or their products."[57] (The example of a culturgen which they seem to prefer is the rejection of incest.) More specifically, they posit the existence of innate epigenetic rules of mental development; cultures evolve as the rules that lead to cultural innovations providing some kind of survival advantage are selected over time, as well the corresponding genes. But it can be objected that this coevolutionary model, "the melding of genetic change with cultural history," to use one of their most felicitous phrases, far from elucidating the origin of the mind, as they claim, amounts to little else than to a mechanization of cultural processes. For it is one thing to liken cultural transmission to genetic processes in order to suggest interesting parallels, and it is another to suggest that there really are cultural forms which are transmitted from generation to generation and whose existence is independent of both the participants in the culture and the observer. As S. Rose *et al.* explains, "interactionism . . . confuses the epistemological plurality of levels of explanation with the ontological assumption that there are really many different and imcompatible types of cause in the real world."[58]

Because it makes the world revolve around a hopelessly subjective observer, the historicist paradigm ends up with a symmetrical rather than a qualitatively distinct epistemology. Before arguing this point any further, however, it is necessary to define the term "historicism," a task which is complicated by the negative, sometimes even nihilistic nature of this movement. Historicism constitutes, above all, a critique of positivism. In this sense it would encompass radical Hegelianism, including Nietzsche, and the whole *Geisteswissenschaften*, but also H. Bergson, the latter Wittgenstein, and even the American pragmatist movement.[59] At the other extreme, F. Hayek and K. Popper's[60] interpretation of historicism as a kind of holistic social engineering that yields predictions about the course of history would exclude many authors (although not all of them) who are characteristic of what is nowadays understood as being an historicist perspective, and would include others (e.g., J.S. Mill or A. Comte) that have very little in common with historicism. Following E.F. Miller,[61] I define historicism as an epistemological position of extreme relativism, characterized by the three following assumptions: i) there can be no direct awareness of pure sense-data, ii) all knowledge is perspectival in character, and iii) truth is relative to time and space. Several philosophical currents share these assumptions without agreeing on their consequences. Among these, the most influential nowadays is probably the new history of science developed by Thomas Kuhn whose ideas find more or less critical echos in the works of I. Lakatos and P. Feyerabend. But, as far as social change is concerned, the historicist tradition is best represented on the one hand by K. Mannheim's sociology of knowledge and, on the other hand, by the various branches of the hermeneutical school.

For Mannheim, all thoughts—with the exception of mathematics and some branches of the natural sciences—somehow reflect the ideological context in which they were formed.[62] In particular, Mannheim criticized Marx for not raising the problem of the social determination of the socialist ideology. In brief, as historical periods succeed each other, so do ideas in an endless chain. Today's truths are tomorrow's errors—the sociology of knowledge has indeed been described as a "sociology of errors." Consequently, human projects are never completely realized and Mannheim, before Daniel Bell and the other proponents of the "end of ideology," anticipated the coming of a positive age when thought would no longer be tinged by ideology or utopia.[63] This curious convergence of historical relativism with positivism points to an underlying epistemological connection (a symmetrical one, as was suggested above). While the former acknowledges the representational character of scientific statements, it shares with the latter a lack of concern for the specification of the conditions of change in the physical or social worlds. The observer is admittedly reintroduced into the scientific discourse, but only after he/she has been rendered mindlessly irrational and unable to reflect upon the substantive content of ideological representations.

The hermeneutic tradition is more difficult to characterize, being more diverse and complex. Many aspects of this tradition can contribute significantly to the construction of a theory of self-organization and will therefore be retained. A useful simplification of the ultimate ends of scientific inquiry consists in contrasting those who seek to explain a sequence of events by identifying the underlying chain of causes and effects (and possibly to predict the occurrence of future events) with those who seek to understand similar events in terms of their relations and meaning within an unfolding scheme oriented towards the accomplishment of a goal.[64] Hermeneutics follows the latter strategy. In the strictest sense, it is the art of Biblical interpretation, and thus we could trace its origin back to medieval theology. The Renaissance humanists applied Biblical interpretative procedures to the classical texts, then from the 18th century onward the scope of the hermeneutic methods has been continually broadened to the point where, as some authors now claim, hermeneutics and philosophy are one and the same.[65] At the same time, the significance of hermeneutics has been confined almost exclusively to the German philosophical tradition. The most frequently discussed contribution of the hermeneutic school to the philosophy of the social sciences are those of Wilhelm Dilthey and Hans-George Gadamer. It is, however, possible to recognize connections between hermeneutic philosophy and other traditions; for instance, in the Anglo-Saxon world, the writings of the latter Wittgenstein and the British analytic philosophy (e.g., P. Winch) have much in common with the German hermeneutic school; both emphasize the importance of language and of the "meaning" of human action and both adopt a "practical" point of view. (In the North American context, one would also have to take into account important works by Charles Taylor, and Michael J. Shapiro.)[66]

In reaction to, but also as a partial concession to the ascent and success of the positivist movement in all fields during the second half of the 19th century, the German historian J.G. Droysen suggested that there is a fundamental difference between history and the natural sciences. While we should expect causal explanations from the latter, the aim of the former is to understand (*Verstehen*) the unique meaning of its subject-matter. Dilthey systematized these ideas and built upon the methodology of *Verstehen* an entire field of human knowledge for which he coined the term *Geisteswissenschaften*, and under which he subsumed everything but the natural sciences. For Dilthey the study of human conduct demands of the observer that he/she reenacts what he/she wishes to understand through a psychological operation of self-transposition into someone else's life. For example, to write about the *indépendantiste* movement in Quebec, one would have to project oneself into the imaginary personality of a committed member of the *Parti Québécois*, at a given time, at a given place, etc. Moreno's experiments with psycho-drama could perhaps be invoked to justify the feasability of such a self-transposition, but the epistemological problem of the communicability of psychological experiences remains the most serious weakness of Dilthey's methodological rules. Either we can only view the world of human action through the eyes of an almost infinite multitude of individuals whose representations will always remain incommensurable, or we submit these representations to some form of external empirical validation, in which case Dilthey's method is not a substitute for, but merely a complement to, positivistic empiricism. The second interpretation, while it is certainly contrary to what Dilthey himself meant by *Verstehen*, has been advanced often by his critics who claim that he mistook the "context of discovery" for the "context of inquiry."[67] This criticism is to a large extent vindicated by Dilthey's psychologistic approach, from which he later tried to dissociate himself.

But the same cannot be said of H.-G. Gadamer,[68] for whom hermeneutics is universal, extending beyond the *Geisteswissenschaften*. It is not simply a method for understanding history, but also an ontological principle that does not lead necessarily to relativistic historicism, because it is not apprehended at the level of the individual (as was Dilthey's operation of *Verstehen*), and because it does not reduce the "traditions" through which we realize their lives to mere ideologies (as Mannheim did). On the contrary, he sees in them the "unity of reason." To analyze the logic of *Verstehen*, Gadamer compares it to the act of translating from a foreign language. Firmly rooted in his own cultural tradition and historical time, the hermeneuticist reconstructs the whole into which the parts of his "text" fit through a series of iterative steps. As the correspondences between his own situation and that under study are progressively elucidated, the interpreter becomes more and more able to disentangle his own values from those of other historical subjects. Moreover, and here are the most original aspects of Gadamer's analysis, the "hermeneutic circle" described above is not only a formal procedure, but it also leads to a substantive reappropriation of one's own tradition through the act of understanding its formative development—which itself results in a further development. ("Understanding is therefore not merely reproductive but also productive.")[69]

Thus in Gadamer's hermeneutics we find a valuable framework for the formulation of an analysis of societal change that would not blind us to the epistemological traps in which the other perspectives discussed so far have fallen. But, as A. Giddens explains (with reference to Apel and Habermas' critiques of Gadamer's work):

> The thesis of the "universality of hermeneutics" could only be sustained if man were wholly transparent to himself, in a world of perfect Hegelian rationality. It is necessary, in fact, to resist the claim to universality, with regard to the explanation of human conduct, of the two major competing traditions of philosophy: hermeneutics and positivism.[70]

The very possibility of a critique, without which it is not even logically feasible to distinguish a dialogue (a concept which is most important for Gadamer) from a monologue, would be ruled out by the a priori rejection of an alternative perspective. There is much to be learned in this regard from Habermas' attempt to complement hermeneutics with systems theory and functionalism which he regards as more or less synonymous terms. In the end, Habermas moves rather far away from orthodox functionalism, but his critique has helped him to work out the hollow mold, so to speak, of his own theory of societal dynamics.

The British cultural anthropologists of the 1920s and 1930s (e.g., B. Malinowski, A.R. Radcliffe-Brown) were the first to introduce the view that social communities could be likened to biological organisms within which certain parts fulfill "some vital function." But, of course, the most significant formulation of the functionalist approach for the study of large scale sociopolitical systems is that of Talcott Parsons, whose "structural-functionalism" also incorporates elements taken from the "theory of action."[71] Indeed, it is this combination of a voluntaristic perspective with a natural-scientific method that renders Parsons' theory so attractive and, ultimately, so disappointing.

For Parsons, a society is a self-sufficient entity. Within it one can identify certain structures, each one of which performs an internal function of integration or mediates the exchanges with the environment through i) the maintenance of controlling patterns, ii) the orientation of the society towards the attainment of specific goals in relation with its environment, and iii) its adaption to the physical environment. The "environment" is a very broad category; it includes the physical environment but also all the other components of the "action system" of which it is itself a part, namely the cultural system, the personality system, and the behavioral organism. The total action system is, in turn, analyzed in terms of the same functional categories—the social system performing at this upper level the function of integration.

Given the importance attached by Parsons to the notion of adaptation, structural-functionalism directly implies the notion of social evolution and implicitly calls for a theory thereof. Indeed Parsons himself, although rather late, has shown that functionalism provides a ready-made frame for the application of concepts and principles derived from evolutionary biology to the description and explanation of societal change.[72] Thus we find in Parsons' work a fully developed conceptualization of change as a variable pattern of relationships between the physical world and the world of human action, as well as of relationships among the components of the latter. These relationships are determined in part by human wants and needs and in part by external necessities.

It may seem that Parsons' approach meets von Foerster's epistemological criteria and that the need for a theory of social evolution is thereby established. On the contrary, my point here is that Parsons' writings on social evolution illustrate the fact that to formulate a theory of social evolution does not suffice to solve the problem of change as it has been posed above. Several authors have attacked Parsons' analysis on social evolution precisely because it tells us so very little about change.[73] And what it conveys about societal dynamics appears to be unacceptably arbitrary, almost idiosyncratic, at any rate definitely ethnocentric. Habermas' critique of Parsons' functionalism makes this very clear. A society's goal states cannot be formulated objectively by an external observer. Therefore, the empirical-analytic paradigm to which Parsons adheres without reservation remains unable "to specify structures essential to the continued existence of the capitalist organization of society and to distinguish them from other elements that can change without the system losing its identity."[74] As McCarthy suggests, Parsons could have adopted a normative-analytic perspective,[75] in which case his presuppositions about what constitutes a vital structural change would not have been presented as nomological statements. But he always refused such an alternative.

So far I have outlined the negative contours of a theoretical interpretation of societal dynamics that would be consistent with the self-organization paradigm. To recapitulate, such an interpretation should not reify the attributes that it ascribes to the objects of study—*a fortiori* the same remark applies to the relationships that are established between those objects. Alternatively, to consider the representations of the world of human action as being essentially subjective and incommensurable, and to regard reason as being encapsulated in the ideological context in which it has emerged would also be inadmissible. On the other hand, functionalism, although it apparently differs from the classical mechanicist paradigm in that it emphasizes the whole rather than the parts, and directionality rather than causality, turns out to be just as arbitrary because it does not acknowledge that social systems possess a self-defined identity. A brief examination of Habermas' treatment of the problem of historical change will help us to sketch the positive contours of a more appropriate epistemological strategy.[76] More specifically, I want to draw attention to Habermas' case for the indispensability of a theory of social evolution.

For Habermas, in order to avoid the arbitrariness inherent in the functionalist approach and to permit the establishment of a critical distance lacking in the hermeneutic perspective,

> . . . what is demanded is a level of analysis at which the *connection* between normative structures and steering problems become palpable. I find this level in a historically oriented analysis of social systems, which permits us to ascertain for a given case the range of tolerance within which the goal values of the system might vary without its continued existence being critically endangered . . . Ranges of variation for structural change obviously can be introduced only within the framework of a theory of social evolution.[77]

Habermas denies that he has constructed such a theory but his writings on this subject already constitute a promising blueprint. His schema is based on the assumption that the evolution of the human species can be understood as a succession of developmental stages whose order is logically reconstructible. For example, neolithic societies organized around kinship structures precede the classical civilizations organized around a state; later the differentiation of the market economy of the feudal societies leads to the development of liberal capitalism which is now going through a transitional (?) stage where the market is recoupled with the political system. This, in itself, is not an original thought; it can easily be traced back to Marx and most Marxist historians since him have taken this perspective for granted. What makes Habermas' approach original is, first, that he differentiates very clearly between the developmental logic that gives meaning to the historical experience of the human species and the contingent factors that actually account for the level of development reached by a given society at a certain time.[78] Patterns of historical development can be reconstructed a posteriori. They show to what extent certain structures have changed relative to others and with what consequences. But they do not explain why, when, and how. The latter questions can be understood within a theory of crises that would, as we have seen, incorporate the concepts of disorder and uncertainty, and relate them to the activities of human actors that create, distort, and manipulate symbols. However, Habermas argues that one can logically construct the space within which these fluctuations occur. More specifically, the paradigm that would be best suited to the formulation of that development logic is the Piagetian theory of learning. And learning takes time; here is the crucial dimension that is lacking in the other paradigms.

This brings us to the second blockbuster. As James Miller notes, it is *prima facie* implausible that "the history of the species passes through universal stages of cognitive and moral development analogous to the stages of individual development observed in advanced societies."[79] And yet Habermas, by projecting on to the societal level all the categories of Piaget's genetic structuralism, including the stages of moral consciousness, claims that the species is actively engaged in a process of moral development. This is a bold hypothesis to make at a time when the Victorian hopes in universal progress have never sounded so hollow.[80] In what follows, I examine the validity of both of these claims (i.e., that social evolution can be understood as a learning process and that Piaget's theory of learning should be used for that task) within the context of what I have called the problem of change, but without referring directly to Habermas' defense of his own position.

In a learning situation one creates one's own identity in an unpredictable way. Errors count often as much as positive steps. Nevertheless, learning is not a random process. As one becomes more aware of the organizing principles of the object of study and of one's own relation to them, a pattern emerges that guides further progress. This equilibration process often obeys a dialectical logic (e.g., the comprehension of the relationship between length, width, and volume involves three phases during which the child first reasons along one dimension, then along the other, before finally combining the two).[81] When learning extends over a long period of time, the patterns vary and learning becomes an evolutionary process (e.g., the stages of cognitive development discovered by Piaget). We have already seen that biological evolution can be likened to a manyfold and highly complex learning process. Therefore, a learning paradigm supposes a reconstruction of reality rather than a transparent account of the actual events. It meets our first epistemological requirement. Thus a learning paradigm that posits a dialectical equilibration of structuring patterns through a logically reconstructable succession of development stages provides us with an adequate conceptualization of evolutionary change.

Historical actors do not confine their learning skills to the acquisition of scientific and technological knowledge. All their activities, including political interactions, provide them with opportunities to learn—or to choose not to learn! Normative considerations contribute to the determination of policy choices as much as the manipulation of empirical conditions. Consequently, if a learning paradigm offers an adequate solution to the problem of change, it ought to be applicable to an analysis of normative structures. Still, the notion of moral development and the transcription of the Piagetian stages into similar categories at the societal level remains *a priori* problematical.

The acid test for a theoretical construction resides in the extent to which an active subject can make meaningful inferences from it. It is my contention that Habermas' evolutionary notions pass this test.

The equilibration of an individual's cognitive structures from a symbiotic to a universalistic stage (through an egocentric stage followed by an objectivistic one)[82] results, for Piaget, from the dynamic matching of internal biological structures with the environment (including socio-linguistic systems) in which they develop. Lawrence Kohlberg and his associates have built a similar scale for the analysis of moral development which Habermas also integrates in his schema;[83] it includes three levels (preconventional, conventional, and postconventional) sub-divided into six stages (from the punishment and obedience orientation to the universal ethical principle orientation). When nothing in the environment can be understood with reference to universalistic criteria, it is probable that few individuals will reach the last stage on Piaget's scale (although some will). Vice versa, when empirical observations of the type made by Piaget and others in Europe and North America establish that most children in these societies reach the last stage of development, we can safely presume the existence of corresponding socio-linguistic structures. But what is needed is a theoretical representation of the relationships (presumably of a feedback type) that develop between the individuals and the collective identity in which they partake. Such is precisely Habermas' concern; his solution consists in positing that i) world views have evolved from the magical-animistic beliefs of the paleolithic societies to "the establishment of universalistic forms of intercourse in the capitalist economy and the modern state,"[84] and that ii) normative structures (i.e., law and morality) have evolved from the preconventional legal regulation of conflict characteristic of neolithic societies to the "general, formal, and rationalized law" and the consequent separation of legality and morality characteristic of modern societies.[85]

The prima facie implausibility of this schema can now be resolved in view of i) its immediate relevance to the concerns of historical subjects actively constructing their individual and collective selves, and ii) its practical and critical import. For example, the alienation experienced by some segments of the population of "developing countries" who, having been individually socialized into modernity, realize the necessity of borrowing secular and universalistic practices and ideas from the Western world, and yet find it impossible to reconcile them with their own culture, vividly illustrates the relevance of a theory explaining the relations developing between cognitive ontogenesis and the social matrix. On the other hand, the possibility of using "the cognitive potential contained in world views for reorganizing action systems"[86] would have a clear practical significance. In conclusion, Habermas' research strategy considerably reduces the arbitrariness inherent in any solution to the problem of change.

However, to analyze the evolution of the species in terms of a developmental logic that includes a moral dimension does not imply that history moves towards the triumph of good over evil. It simply means that we can—but will not necessarily—become more and more conscious of, and act upon, the hierarchy of values sustaining political life. To achieve this, planning models constructed for the purpose of clarifying the debate among rational political actors about societal priorities can be useful. But such models risk being treated as pseudo-scientific justifications for misguided policy choices, for instance by demonstrating the necessity to "adapt" to reified changes in some aspects of the system (e.g., technological advance), unless they are grounded in a theory of social evolution derived from principles similar to those outlined by Habermas.

The Old and the New Evolutionists: Unconvincing Answers

Now that the use of an evolutionary perspective in social research has been justified, I still have to explicate in more detail what it entails. For this purpose, I review the evolutionary schemata of some important social theorists, beginning with Herbert Spencer. My purpose here is not to articulate a comprehensive analysis of all the writings of all evolutionists, but rather to zero in on the most significant theoretical and empirical problems raised by evolutionary models of social change. Besides, I am only interested in general theories at this particular stage; specifically political aspects of social evolution (e.g., the emergence of the state) are dealt with in subsequent chapters.

Social Darwinism is generally condemned out of hand by contemporary authors. There is no point in trying to rehabilitate this much-maligned movement for it is indeed bad science, at least from the vantage point of modern biology—a case in point is Herbert Spencer's uncritical adherence to the Lamarckian paradigm—and bad social theory, because it rests on blindly accepted ideological presuppositions. But it is more useful to criticize the actual content of a theoretical aberration than to poke a straw-man.

There are two kinds of social Darwinism.[87] The first variety was founded by Herbert Spencer and was continued in the United States by W.G. Sumner and A.G. Keller; it is strongly individualistic. The second and less well-known movement originated in Germany and Austria; it is, by contrast, a "theory of group conflicts and social determinism which was outspokenly contemptuous of the value of human life."[88] However, Matson argues that both movements share a common mechanistic world view insofar as they posit the existence of a moving force running its course according to the universal laws of nature (but working either through individuals or through groups).[89]

Spencer remains the most interesting of the earlier evolutionists because he was primarily a political theorist. Placed in the context of his life-long defense of "negative" liberalism—i.e., the minimization of coercion by the state—his theory of social evolution is but a means of scientifically demonstrating the validity of his position regarding natural rights and the state of perfect competition.[90] Spencer's writings on evolution deserve our attention for at least two reasons. First, it exemplifies the ambiguity and complexity (meaning here conceptual richness) of the notion of evolution. What Spencer saw as the necessity of evolutionary progress can sometimes be reinterpreted, using his own concepts, as the uncertainty and indeterminacy of systemic changes.[91] Second, there is a clear, although indirect, link between Spencer and the neo-evolutionists, and with Talcott Parsons in particular.

The three "laws" of evolution proposed by Spencer imply something that is not as consistent with his own Newtonian world-view as he himself thought they were. "The Law of the Instability of the Homogeneous" states that "homogeneity is a condition of endemic instability, perpetually under attack from unequal and conflicting forces which ultimately bring about its *dissolution*, resulting in *a progressively intensifying heterogeneity*."[92] So Spencer attributes an incoherent character to homogeneity and, conversely, a coherent character to heterogeneity. Entropic chaos fits Spencer's view perfectly but, as it stands, Spencer's first law contradicts the Second Law of Thermodynamics since the latter posits a reverse sequence of events. Had Spencer been able to propose an explanation for this contradiction, his intuition regarding the coherence of heterogeneity and the dynamic dissolution and reconstitution of order from one level of complexity to the next, which anticipates the "order from noise" principle, would have allowed him to make an important theoretical breakthrough.

The "Law of the Multiplicity of Effects" ("Every active force produces more than one change—every cause produces more than one effect")[93] and the "Law of Segregation" ("Unlike units of an aggregate tend to separate and like units to cluster, resulting in a progressively sharpening structural differentiation")[94] could just as well describe the effects of, respectively, positive and negative feedback. My point is that Spencer understood the complexity inherent in the evolutionary process but did not explicitly thematize it as an epistemological problem. He hit upon some far-reaching ideas (e.g., the three "laws") without being able to realize the extent to which they were heterodox. Therefore, he failed to break away from the mechanistic assumptions of 19th century science. Moreover, Spencer used his theory of evolution to prove that it is impossible to compute the costs and benefits of alternative societal choices. This, in turn, justified in his view a state of competition among enterprising but blind individuals whose ontological status evokes that of atoms in classical physics.[95] The subtle interactions between populations and their environment outlined by C.H. Waddington do not find their place in Spencer's theory; like Lamarck, he conceived the environment as a source of functional needs that must be fulfilled but never as a challenge.

The most contentious, even pernicious, aspect of Spencer's work is his contempt for history.[96] He did not hesitate to force singular historical events back into his evolutionary schema built from a priori principles (rather than reconstructed a posteriori) and articulated around causal relations (rather than following an interpretative method). Some contemporary authors come dangerously close to this methodological aberration.

At one point Spencer believed, like most 19th century ideologues, in the ineluctability of universal progress. He later came to acknowldge the possibility of divergent evolution. Neo-evolutionist authors, such as N. Bellah, S.N. Eisenstadt, and above all, T. Parsons, never reduced social evolution to a strictly unilinear development. Nevertheless, both Spencer and Parsons share the view that the course of evolution is determined by functional prerequisites and law-like connections between the progressively more differentiated components of social systems. I have argued already that this vitiates Parsons' whole theoretical schema, but it is now time to examine it more closely.

There are some valuable insights to be gained from Parsons' transposition to the level of sociological analysis of the biological notion of "evolutionary universals" (e.g., vision) which echoes Waddington's "archetypes." However, the relevance of these ideas hinges on Parsons' overemphasis of the explanatory power of the ill-defined concept of adaptation.

> Among change processes, the type most important to the evolutionary perspective is the *enhancement of adaptive capacity*, either within the society originating a new type of structure or, through cultural diffusion and the involvement of other factors in combination with the new type of structure, within other societies and perhaps at later periods.[97]

But what does he mean by "adaptive capacity"? Sometimes the term is used in a quite specific sense, i.e., the capacity to adjust to (or even to transform) the physical environment. The economy performs the task of adapting the social system to the physical environment and the "behavioral organism" performs a similar task with respect to the entire action system. But generally the nature of the environment to which the action system or its subsystem adapts remains unspecified. It thus appears that *all* his functional categories are particular forms of adaptation and that all structures perform an adaptive function. Parsons seems to suggest that adaptation allows each structure in the social system to relate better to its corresponding structure in the entire action system, but then what does the action system adapt to? Uncertainty in the environment (which must be reduced)—that would be Parsons' answer. But, as A. Giddens observes, "since 'uncertainty' is nowhere defined, the thesis is conceptually so diffuse as to be virtually useless."[98] Moreover, uncertainty is not something that necessarily must be reduced through the process of adaptation, since it has been stressed already that change in complex systems often occurs in crisis situation. As evolution usually results in an increase in complexity, uncertainty about the present or future state of internal and/or external structures is more properly understood as a measure of the degrees of freedom characterizing the strategies of the self-organizing biological or social entities comprising a given eco-system than as a threat to the survival of these entities. For example, technological innovations (e.g., recombinant DNA) that accompany or stimulate important social changes may reduce the uncertainty inherent in some existing relationship with other components of the bio-societal ecosystem (e.g., food production) but, at the same time, unsettle a number of other structures (e.g., the economy of rural regions, the ecological balance, and even our ethical notions about acceptable risks, the sanctity of life, etc.).

To the extent that Parsons regards social change as being oriented towards the gratification of individual needs, and more importantly towards the cumulative development of cultural traditions, one could conclude that evolution points to the creation of a humanized world. However, what Parsons means by the gratification of human needs is simply the realization of the rather narrowly defined "need disposition" of the personality. Furthermore, he suggests that adaptive pressures that are largely independent from the will and aspirations of human actors bring about a complexification of society and a consequent rationalization of social interactions. In brief, Parsons approaches the subject of social evolution from an essentially deterministic point of view.

This appears clearly in the way he uses the potentially meaningful concept of "evolutionary universals."[99] The formation of thresholds characterizes the dynamics of complex systems. They serve to structure the patterns of development (e.g., Waddington's chreods or Parsons' evolutionary universals) ensuring the reproducibility of differences. In this respect, Waddington distinguishes the "archetypical" from the usual type of epigenetic chreod. The former are "not created by natural selection but evolve as a consequence of some 'internal necessity',"[100] whereas the latter are created by natural selection, in the sense that they emerge from an organism's interactions with its environment. This distinction is fundamental because it prohibits the direct application of the organizing principle at work in archetypical chreods (e.g., the insect form) to the development of actual populations. A wide range of intermediary variables must be taken into account. Some of these variables can, as we have seen, be interpreted as the result of unpredictable "choices." At the biological level there are obviously many other types of more objective determinants of evolutionary changes. But at the societal level the relative importance of objective and subjective factors is reversed. To conclude, the notion of evolutionary universal (or archetype), borrowed by Parsons from biology, does not mean that evolution results from a uniform process of functional differentiation, that is to say it is not necessarily a-historical. Contrary to Habermas, Parsons fails to distinguish consistently between the archetypical patterns and the actual epigenetic processes. He is aware of the difference, but he does not consider it as problematical. He acknowledges that the institutionalization of the evolutionary universals and, in particular, of the four complexes characteristic of modern society[101] is uneven, and yet he treats these discrepancies as mere details, claiming that empirical analysis would later fill in the gap. However, it must be emphasized that the emergence of a specific form of political organization within the logical space of an evolutionary universal constitutes an important problem that requires the formulation of appropriate concepts and models. The theory of self-organization, Waddington's epigenetic paradigm and the emerging "crisology" that were discussed previously can be combined to take up this challenge without cutting off the diachronic developmental processes from the observer's view.

Parsons' characterization of British Common Law as *the* ideal type of a generalized universal legal system exemplifies the kind of mystifying conclusions that his inability to separate the logical from the historical leads him to draw. It may be true that the Common Law protects individual freedoms better than legal systems derived from the Roman law tradition. On the other hand, the sometimes labyrinthian procedural provisions of the Common Law render it less universalistic, at any rate less rational than Roman Law. This explains in part the fact that most non-English-speaking countries have adopted the Roman Law tradition (although some of these versions are quite heterodox, e.g., the "socialist law" of Eastern European countries). There is little doubt that a universalistic legal system represents an important building-block of societal evolution. Societies that fail to "hit upon" such an organizing principle cannot escape the confines of their niches, where they risk extinction. However, the Common Law is clearly a unique historical development. It must be understood as a fundamental component of the life-world of the Anglo-American civilization.

This is not meant to imply that the functionalist perspective on the subject of social evolution is wholly mistaken. An exclusive preoccupation with world views abstracted from their structural setting would preclude the critical analysis of their content. This is precisely what Habermas finds wrong with Gadamer's hermeneutical interpretation of the self-formative process of historical traditions.[102] The problem is how to reformulate these two approaches into an integrated representation of the history of the species. I have explained already why I regard Habermas' strategy as being very promising; however, I take issue with some aspects of his tactics.

As it has already been explained, a theory of social evolution constitutes for Habermas a *sine qua non* for the formulation of a social theory with a practical intent. He regards Marx's historical materialism as an imperfect approximation of such a theory—and not simply as a mere heuristic device, even though Marx did not elaborate on the theoretical foundations of his method, preferring instead to deal directly with particular historical situations.[103] Habermas is not satisfied with some of Marx's assumptions—e.g., the importance attached by Marx to the notion of social labor—and, consequently, undertakes an overall reconstruction of historical materialism borrowing several elements from Piaget's genetic structuralism.

My critique focusses first on the strictures that even a reconstructed historical materialism imposes upon Habermas in his effort to synthesize a wealth of recent anthropological studies, and second on his refusal to follow Piaget on the bridge that the latter has built between the biological and the socio-psychological worlds.

Drawing from the works of authors such as E.W. Count, and interestingly, E. Morin,[104] Habermas concludes:

> The Marxian concept of social labor is suitable for delimiting the mode of life of the hominids from that of the primates; but it does not capture the specifically human reproduction of life.[105]

That specificity must be located in the "familiarization" of the male, itself connected with the development of cooperative hunting practices. The division of labor between the hunting males and the females who cared for the young resulted from the adaptive move that some chimpanzee-like primates made when they left the protective forest for the open savannah. This initial "choice" triggered a series of positive feedback interactions affecting factors like i) biological roles (which became more specialized), ii) the criteria upon which social hierarchies are established (which became more equilitarian), iii) the use of tools, and iv) the variety of the communications patterns (in Ashby's sense of the term) resulting in the development of "protolanguages." Thus, in hominid groups, the structures of labor and language underwent profound qualitative changes; and these changes allowed morphologically innovative, but otherwise extremely vulnerable creatures to thrive in their new environment, and accelerated the growth of their neo-cortex. From this new evidence we can conclude that "labor and language are older than man and [specifically human] societies."[106] Unfortunately, Habermas limits the import of this far-reaching observation to the proposition that the concept of social labor "cuts too deeply into the evolutionary scale"[107] to be used as the primary factor explaining historical development, and chooses not to pursue the inquiry into the transition between biological and societal evolution any further. But more will be said about this later.

With the integration of the adult male into a family structure there appeared for the first time a set of societal norms designed for the regulation of a sociolinguistic system, as opposed to a sociobiological one. These norms resolved the contradictions that had developed between the male and female subsystem. But they also produced new elements of disorder—a specifically human kind of disorder, e.g., the double identity of the child who now enters into imitative/antagonistic relations with his father, or the cooperative/competitive interactions that developed between different hunting bands as a result of the practice of exogamy, etc.[108] Therefore, the origins of a specifically human mode of life are not to be found in the transition to new forces of production, but in their patterning within the context of a new mode of social integration. As Habermas concludes:

> Rules of communicative action, that is, inter-subjectively valid and ritually secured norms of action, cannot be reduced to rules of instrumental or strategic action. Production and socialization, social labor and care for the young, are equally important for the reproduction of the species; thus the familial social structure, which controls both—the integration of external as well as of internal nature—is fundamental.[109]

What is true of the origins of man should a fortiori be true of human history. Yet, in trying to rescue as much of Marx's work as he can from the breakdown of the simplistic dogma of the primacy of narrowly defined economic infrastructures (for which Marx is admittedly only indirectly responsible) under the attacks of the new anthropology, Habermas retrogrades to some rather contrived theoretical positions. While he is more concerned with the development of holistically defined "social formations" than with the dynamics of the forces of production per se, Habermas nevertheless claims that the latter are part of the basic domain of society. Evolutionary innovations can be made only at this level; however, they do not directly bring about an evolutionary step. New knowledge needs to be implemented to become an instrument of change. "But [Habermas adds] only when a new institutional framework has emerged can the as-yet unresolved system problems be treated with the help of the accumulated cognitive potential."[110] Far from being the effect of a qualitative transformation in the forces of production, a new form of social integration often causes a quantic leap in the development of these forces by facilitating the implementation of a yet untried cognitive potential. And how are these new institutional frameworks evolved? They are consciously designed by the social actors who, as we have seen already, can learn not only "in the dimension of technically usefull knowledge crucial to the development of productive forces but also in the dimension of moral-practical consciousness decisive for structures of interaction."[111]

What this means is that, for Habermas, the superstructure, or modes of social integration, are not mere passive reflections of the infra-structure,[112] but operate a creative mediation between the forces of production and social development. As a theoretical proposition, this characterization is certainly instructive. However, in Habermas' text we can also discern a subtle, and in my view quite unacceptable, transposition of this analytical distinction from the level of discourse to that of reality. This is due to his refusal to break explicitly with Marx's ontological materialism.

It seems that, for Habermas, the superstructure is not only secondary in an analytical sense but also secondary in a real sense, to wit:

> the rules of communicative action [the concrete realization of which form the superstructure of society] do develop in *reaction* to changes in the domain of instrumental and strategic action.[113]

The qualification of this statement to the effect that "in doing so they follow *their own logic*" does not fundamentally alter its meaning.

By "secondary in an analytical sense" I mean a concept implying a self-referential construction. By "construction," I refer to the logical operations by means of which a concept is formulated; and by "self-referential," I mean that the concept enters into its own definition. The production and reproduction of the institutional framework of social integration, that is to say politics broadly defined, should be understood as a second-order process in a similar sense. Politics supposes first-order processes, e.g., the dynamics of the "productive forces," insofar as it is superimposed upon them. For example, the bureaucratized states of the archaic civilizations (e.g., Mesopotamia or Egypt) modified the ways in which their agricultural economies functioned by organizing the redistribution process with remarkable efficiency. In the case of Egypt, this was done by centralizing the distributive system around the person of the pharaoh who made "allotments to his officials who in turn rewarded their retainers and so down the social scale."[114] In this instance the political system performed a second-order distribution, that is to say a distribution of distribution; but both the distribution (from producers to consumers) and the distribution of distribution (from official to official) did partake of the *same* reality (i.e., the ability to participate in conventionally validated social relations).

To speak of a "reaction" (see Habermas' quote above) seems to imply not only a subjectively created (i.e., both by the observer and the societal actor) but also a real, objective relation of functional dependence, as the physical connotation of this term would indicate. Yet, as Morin puts it, "the economy is a base of culture [this is an analytical proposition], and not the 'other' base upon which culture rests and develops."[115]

Despite Habermas' repeated and well-taken caveats about the risks involved in directly applying his developmental logic to the explanation of actual historical events whose contingent character would thereby be ignored, he at times comes dangerously close to making such a fallacious move. His adherence to even a modified version of the infrastructure-superstructure schemata does not allow him to break once and for all with linear causality. As long as it is permissible, "in the final analysis," to speak of unequivocally determining factors (e.g., the innovations required for the occurrence of an "evolutionary step" can be located in a reified "base" of society), one cannot preclude the possibility of reducing what is presented as a dialectical and *sui generis* learning process to a straightforward causal explanation admitting of unmediated empirical applications.

Of course, political institutions function within certain limits set by some very real factors (e.g., size of the population, access to physical resources, technological advance, etc.). Moreover, societal evolution cannot proceed in all directions at once; in the same way that it is highly doubtful that biological evolution could ever produce a feathered fish, societal evolution could hardly realize the combination of Jeffersonian democracy with the nomadic life-style of the Berbers. But in this respect it is more appropriate to speak of contextuating influences circumscribing both the "base" (Habermas), or infrastructure, and the "rules of communicative action" (Habermas), or superstructure.

As far as the question of the relation between nature and culture is concerned, Habermas' lack of interest in evolutionary biology has shielded him from the most suggestive epistemological implications of the epigenetic paradigm. For instance, Waddington's interpretation of the relationship developing between the genotype—in an illusory way, the "base" of the evolutionary process—and the phenotype clearly illustrates the kind of circular process to which I alluded above; as it has been explained already, the phenotype operates second-order transformations that bring the productive processes regulated by the phenotype to bear upon themselves. Similarly, P. Corning's detailed account of the role played by functional synergism at all levels of the evolutionary process, including socio-cultural evolution, reveals the futility of any attempt at maintaining a dogmatic distinction between a "base" and a "superstructure" (see below). Furthermore, Habermas could have tried to ground his developmental logic of world-views and morality in Piaget's reflections about the correspondence obtaining between organic and cognitive regulations. Space lacks here to elaborate this point, but it seems that, in solving the problem of the interdependence of collective representations and individuals' learning abilities, Habermas leaves the question of the origin of the latter unanswered.

Toward a Solution?

The three authors whose works are reviewed below, namely Y. Barel, P. Corning and E. Morin, have attempted somewhat more successfully than any of the authors discussed above to bring together evolutionary biology, anthropology and social theory with a view to underline converging themes and approaches. Although their works differ in many ways, they share some fundamental assumptions that I also consider to be both valid and necessary. First, they all emphasize the quasi-intentional character of biological reproduction. Hence, for them, the concept of adaptation does not mean a kind of rigid determinism but rather an interplay of evolutionary strategies bounded by an ecosystem which is itself open to change. Second, they neither try to reduce socio-cultural phenomena to biological ones, nor do they seek to justify social practices in terms of some vague evolutionary "laws." What they attempt to do is to show that autonomous, self-directed complex systems exist in both the natural and the social spheres and to understand invariance and change in these systems by treating the specificity of each sphere as a problem to be investigated rather than as an ontological a priori.

As I have indicated already, Barel has done much to substantiate the intuitive realization that biological and social concepts overlap (e.g., development, growth, reproduction). The reason being not that social theorists have followed the lead of the life sciences, but rather that both the life sciences and the social sciences have drawn from the same conceptual well; in recent years, information theory and cybernetics have provided more formalized definitions of these common notions.[116] Hence, a thorough investigation of the conceptual linkages between the life sciences and the social sciences can serve to sharpen our understanding of the specificity of the human and societal aspects of the universal dialectic of invariance and change. (To come back to what I called earlier the problem of change, we reduce the arbitrariness inherent in any specification of what changes or what evolves by clarifying what we understand to be changes brought about by consciously applied, political means, changes brought about by our involvement in more diffuse cultural and somatic structures, and changes in the ecosystem.) Barel begins by stressing that cultures are much more open than genetic pools in that they can borrow elements from each other and that there is almost no limit to these forms of "hybridization." (Yet this does not mean that nature and history are antithetical phenomena; it means rather that as one moves along the uninterrupted chain extending through all living systems, progressive transformations bring about qualitative differences within a unified field of organizational principles.) His most original contribution stems from his analysis of the "unbinding" of social reproduction vis-a-vis biological reproduction. The first aspect of this tendency consists in that the means of production acquire an existence separate from that of the biological organism and become part of a self-reproducing system.[117] However, Barel does not suggest that the reproduction of the means of production is *the* basis of societal reproduction. The second aspect of unbinding is the symbolization of communication and the reproduction of symbolic meaning through a variety of modes of discourse (including political communication).[118] The third aspect is the memorization of cultural norms by autonomous individuals who never reproduce what they have learned in an identical manner.[119] Finally, societal reproduction depends on the combination of a multiplicity of partial reproductions (e.g., material, symbolic, etc.) but the interaction between these various processes is influenced by the political process. The latter consists precisely in establishing hierarchical relations of domination/subordination controlled by a complex system that includes, at least in the modern age, state institutions and class formations.[120] (By contrast, biological reproduction is a more diffuse process.) In other words, societal reproduction leaves even more room for change than biological reproduction and, more importantly, there is no authoritative way of interpreting these changes as they result largely from symbolic creations; however, political life can be understood as an attempt by dominant social actors to impose their own interpretation of these on-going changes.

While Barel's work can best be described as a perspective or a theoretical framework, Peter Corning has worked out an elaborate theory from which he derives a series of empirically testable hypotheses. His most original argument is that "the Darwin of the social sciences is Darwin";[121] that is, Darwin's own theorizing on man and society rather than Spencer's misrepresentation of their import for social theory ought to be the starting point of any reflection on evolutionary social change. The reason being that Darwin never adhered to any kind of finalist worldview; consequently, his theory of natural selection can hardly serve to justify a preconceived notion of "progress." Already incipient in Darwin's theory is the notion of internal directedness or "teleonomy." Corning shows that the causal processes that are responsible for the "progressive" character of evolution can best be explained by reference to the concept of functional synergism (i.e., "systemic properties that cannot be predicted from an understanding of the components").[122] At the social level, synergestic combinations become even more probable and complex. (An example provided by Corning is that of Japan's economic success which "is the result of a synergestic combination of economic, social, and 'political' factors . . . It derives from the combination of a functionally compatible set of cultural values and traditions, lessons that were derived from Japan's historical experience, advantages in the . . . market place, and a highly intelligent set of societal development strategies that its leaders have assiduously pursued for many years."[123]

Not all synergestic combinations, however, have the same chance of being adopted. First, they must not endanger the survival of the biological or societal systems involved. While Corning admits that reproductive fitness is not useful for analyzing adaptation in human societies, he insists that "bioeconomic measures" of the satisfaction of "primary needs" (e.g., safety, nutrition, sex and reproduction, etc.) can be perfected for the purpose of evaluating social development strategies—this may appear to be somewhat naive, but Corning's argument can hardly be dismissed today in the context of the catatrophic and tragic maladaptation of the socio-economic systems of most African countries, as well as of so many marginal groups in the more advanced economies. More importantly though, societal systems are steered by political processes that are analyzable in cybernetic terms. That is, the choice among synergetic combination is a political process involving the formulation of goals and the control over their implementation—sometimes, these goals may even be inconsistent with the satisfaction of at least some basic needs but, in the long term, failure to sustain a developmental effort can be traced to political disintegration.

Now there are several aspects of Corning's theory which do not fit within the overall epistemological strategy proposed in this book. For example, the criticisms that I have directed at functional explanations in the social sciences apply to Corning's functionalist definitions of many of his key concepts (e.g., synergism). Even more seriously flawed is his understanding of cybernetic processes as being primarily, if not exclusively oriented toward the achievement of some kind of internal balance (even though he uses the term "self-organizing system" repeatedly)—in the previous chapters I have explained why this is incorrect, and in the subsequent chapters I intend to take a closer look at some typical political problems from the standpoint of the self-organization paradigm. But Corning's research program, as opposed to the details of his theory, constitutes in my view an important contribution to social theory. His very convincing demonstration of the fact that the directionality of biological evolution does not need to be predicated on any kind of othogenetic assumption is a solid foundation on which one can base a similar argument for the dispensability of monocausal theories of social dynamics: it is precisely because societal relations are complex and unpredictable that they can become organized into competing strategies by purposeful individual and collective actors. Although perhaps less convincing, his claim that the adaptation of societal structures must be interpreted in terms of the satisfaction of some basic needs is a useful reminder that social evolution is not completely divorced from biological evolution. Finally, his detailed analysis of the role played by political communication in social change, his insistence on what he calls "the centrality of politics" adds credence to the position that I defend from here on.

Edgar Morin's ambitious reconstruction of the conceptual apparatus of biology in accordance with the subjectivist epistemology that he feels best corresponds to our experience of life comes closest to my own understanding of evolutionary processes. For Morin the phenomenology of living systems, from cells to human societies, consists in a web of relations of inclusion and exclusion among which two more closely knit networks of specialized interactions stand out; one is centered on the ecosystem in which particular organisms live and reproduce, the other is centered on the autonomous organism itself and constitutes its own sphere of autonomy.[124] In other words, the immense complexity of life consists in a series of partially contradictory and partially re-inforcing organizational processes producing, on the one hand, a cycle of selective, adaptive and integrative interactions and, on the other hand, autonomous, autopoietic unities. But Morin is not merely opposing two polarities (i.e., the environment and the individual); rather his self-organizationalist perspective reveals interesting linkages between the notions of ecological balance and epigenetic development. I deal at some length with his provocative analysis of the identity and sense of self which characterizes every living creature further on (see Chapter 8). On the subject of evolution, which is a phenomenon that is approached usually from the standpoint of ecosystemic processes (e.g., population dynamics, kin-selection, competition among species, etc.), Morin affirms that "the self-organization of living systems supposes and necessitates dependency with respect to the eco-organization within which they are included; the latter, however, supposes and necessitates self-organizing systems which constitute its biocoenosis." He adds that self-organization and ecology are concepts that "cannot be thought of, nor dealt with separately from each other."[125]

Morin pursues the theme of the paradoxical complementarity between self-centered autopoietic processes and systemic integration into a higher order of processes by examining the evolutionary transition from unicellular organisms through polycellular organisms to animal and human societies. Thus he suggests that, because mammalian societies are formed by communicative interactions among sexually differentiated individuals possessing complex nervous systems, "[the existence of a] society supposes not only individualized members (which is already the case with [polycellular] organisms) but very highly individualized members."[126] That is to say that the traditional opposition between the ideas of individual and society is unwarranted. On the question of the specificity of human societies, Morin implies that they reach a genuine identity based on cultural structures, whereas animal societies remain "uncompleted." In the modern age, this completed integration is typified by the nation-state (see Chapter 7 where I elaborate on this idea)—but there is no reason to think that this kind of integration cannot be transcended.

Concluding Remarks

Of all the problems that students of politics are faced with, the most difficult is that of distinguishing between invariance and change and of understanding the logic of societal change which forms the background of contingent, unpredictable events. I have attempted to show that evolutionary biology provides some insights into this problem. Obviously, the experimental fact that human actors can choose between alternative modes of action constitutes a valid point of departure. However, it is too crude an observation to be theoretically meaningful since many non-human living beings behave in ways that are not determined entirely by innate factors. Indeed, as my survey of evolutionary biology has shown, evolution is a goal-directed process that reflects the interplay of the strategies adopted by competing and/or co-operating living systems. Hence, it is incumbent on social scientists to be very explicit about the specifically human and social attributes of whatever type of behavior they are interested in. The problem of making sense of societal change is further complicated by the absence of a reference point situated outside of the world inhabited by the observer himself or herself, and in particular, outside of the sphere of linguistic practices.[127] I have suggested that an evolutionary vision offers some guarantees against arbitrariness. As many of the authors reviewed in the preceding pages have stressed, human evolution is oriented toward the emergence and development of *political* processes of integration/disintegration and re-organization. It follows that the next task is to examine the various dimensions and implications of political life.

Notes

1. Peter Corning, *The Synergism Hypothesis: A Theory of Progressive Evolution* (New York: McGraw-Hill, 1983).

2. H.R. Maturana and F.J. Varela, "Autopoiesis: The Organization of the Living," in H.R. Maturana and F.J. Varela, *Autopoiesis and Cognition* (Boston: D. Reidel, 1980), p. 107.

3. See his *Tools for Thought* (New York: Basic Books, pp. 16ff); it would also be inane to associate Whitehead with such a view, since he so vigorously denounced the "fallacy of misplaced concreteness"; see his *Science and the Modern World* (New York: The Free Press, 1967), p. 51.

4. See C.H. Waddington, "The Practical Consequences of Meta-physical Belief, on a Biologist's Work. An Autobiographical Note," in C.H. Waddington (ed.), *Towards a Theoretical Biology*, Vol. 2 (Edinburgh: Edinburgh University Press, 1969), p. 75.

5. See *La méthode*, Vol. 1, pp. 155–181.

6. In E. Morin's words: " . . . artificial machines are, of course, capable of generating energy, of developing an informational competence and organizational autonomy. But they have not succeeded in developing organizational generativity.

They are only able to develop phenomenal organization that produces outputs, but not generative organization that produces its own means of production itself. . . . [Thus] our artificial machines should not really be considered as machines but as some kind of prosthesis in the social megamachine (ibid., p. 170, my translation)."

7. F. Jacob, *The Logic of Life* (New York: Pantheon Books, 1973), p. 74.

8. Quoted by C.H. Waddington in *The Strategy of the Genes* (London: Allen and Unwin, 1957), p. 3.

9. *Autopoiesis* (1980), Chapter 3, *in passim*.

10. See K. Boulding, *Ecodynamics* (Beverly Hills: Sage, 1978), Chapter 5; incidentally, Boulding's account of evolutionary biology is helpful, although he makes two mistakes (on the number of chromosomes in man, p. 102, and on the classification of the Australian dingo, p. 113).

11. See E. Mayr, *The Growth of Biological Thought: Diversity, Evolution and Inheritance* (Cambridge, Mass.: Harvard University Press, 1982), pp. 566–570.

12. Because of the confusion surrounding the "question of reduction," the term "reductionism" should never be used without a qualifying statement. Dobzhansky *et al.*, clarify this issue by distinguishing three kinds of reductionism; see T. Dobzhansky, F. Ayala, J.L. Stebbins, and J.W. Valentine, *Evolution* (San Francisco: W.H. Freeman, 1977), pp. 487–495. "Ontological reductionism," as far as biology is concerned, asserts that living organisms are "ultimately made up of the same atoms that make up inorganic matter, and nothing more" (ibid., p. 488), which is not to say, however, that they are *nothing but* atoms and molecules. I concur with the authors who see ontological reductionism as a prerequisite for the scientific study of living systems (with the *caveat* that atoms are themselves rather fuzzy entities, being both matter and energy). "Methodological reductionism in its extreme form claims that biological research should be conducted only at the level of physiochemical parts and processes" (ibid., p. 490); this can only lead to a dead-end (the authors also reject this claim but are perhaps too eager to endorse a "moderate methodological reductionism"). To define epistemological reductionism, the authors refer to E. Nagel's formulation of the problem. A science can be reduced to another if, and only if "the laws and theories of the secondary science can be derived from the laws and theories of the primary science" (ibid., p. 493), i.e., the conditions of derivability, and also if, and only if, the primary science provides the premises of a deductive argument the conclusions of which form the secondary science, i.e., the conditions of connectability. The authors claim that "epistemological reduction of biology to the physical sciences is not possible at present" (ibid., p. 949).

13. A more explicitly cybernetic account of this model is given by Miller: "The information which specifies the organization of an organism constitutes a *message* which is transmitted around a *feedback loop* (the life cycle). From time to time *noise* (mutations) arises in the message, which diminishes the *gain* of the message around this loop (mutations decrease the fertility of the species). Those messages that pass through the filter (environment) with a gain equal to or greater than one, produce *positive feedback* (self-sustaining reproduction of the species). Noisy messages, or unfavorable mutation, which cannot pass through the filter, are bred out of the species after passing over the feedback loop a number of times. Eventually the character of the message is determined primarily by the filter." J.G. Miller, *Living Systems* (New York: McGraw-Hill, 1978), p. 77.

14. See his *Chance and Necessity* (New York: Alfred A. Knopf, 1971, transl. by A. Wainhouse).

15. Conflicting views about this question are presented in T. Dobzhansky, *et al.*, *Evolution*, pp. 5–6, and in Pierre-P. Grassé, *Evolution of Living Organisms: Evidence for a New Theory of Transformation* (New York: Academic Press, 1977), Chapter IV and *in passim*.

16. I know of at least two authors who use the term "freedom" to describe the phenomenology of life: H. Jonas (in *The Phenomenon of Life*, New York: Delta Books, 1969, p. 3) and F.W. Matson (in *The Broken Image: Man, Science, and Society*, New York: George Braziller, 1964, Chapter V).

17. C.H. Waddington, "Sketch of the Second Sarbelloni Symposium," in C.H. Waddington (ed.), *Theoretical Biology*, pp. 3–4.

18. See C.H. Waddington, "The Basic Ideas of Biology" in his *The Evolution of an Evolutionist* (Edinburgh University Press, 1975), p. 217.

19. See *Chance*, pp. 94–95.

20. C.H. Waddington, "How Much is Evolution Affected by Chance and Necessity?" in J. Lewis (ed.), *Beyond Chance and Necessity* (London: The Garnstone Press, 1974), p. 97.

21. For Waddington, this adaptation is not due to the "non-genetic plasticity of the phenotype" (Mayr) which, according to some neo-Darwinists could produce a non-hereditary response; only when a random mutation throws up a new allele would this adaptive modification become an inherited one. On the contrary, our author argues that the genotype is not irrelevant to the plasticity of the phenotype, even if it does not account entirely for it. See *The Evolution of an Evolutionist*, p. 89.

22. Ibid., pp. 100–101.

23. The cybernetization of the epigenetic paradigm envisaged by Waddington would involve a much more complex model than that proposed by Miller (see note 13).

24. Ibid., p. 97.

25. My account of Whyte's ideas is derived from E. Jantsch's rendering of them in his *Design for Evolution*, pp. 40–41. For a more complete exposition see L.L. Whyte, *Internal Factors in Evolution* (New York: George Braziller, 1965).

26. A. Koestler, "Beyond Atomism and Holism," in J. Lewis (ed.), *op. cit.*, p. 62.

27. A. Koestler, *The Ghost in the Machine* (Chicago: Gateway, 1971), pp. 146–147. Koestler saw a manifestation of these laws in the archetypical nature of biological form first intuited by Goethe and further articulated by D'Arcy Thompson; on this subject Waddington wrote: "[the small-scale processes of natural selection] have produced a restricted number of 'archetypes' . . . which are flexible enough to become adapted to almost any system of life yet have sufficient inherent stability to do so without losing their inherent stability." (*The Evolution of an Evolutionist*, p. 200).

28. See ibid., Chapter XI.

29. Ibid., p. 152.

30. It is interesting to note that a number of far-reaching exceptions to the Mendelian dogma have recently been discovered. For instance, we know now that

the genetic material of bacteria includes some particles called "episomes" and located within or without their chromosomes which do not derive from a mutation but from external causes (see J.P. Grassé, *Evolution of Living Organisms*, p. 257). Also, D. Baltimore has confirmed experimentally A. Temin's hypothesis regarding the possibility that some viruses reproduce by altering the DNA of their host cells (a more recent experimental confirmation of Temin's discovery has been achieved by E.J. Steele of the Ontario Cancer Institute).

31. *The Logic of Life*, p. 302.

32. *Synergism*, pp. 4–5, see also p. 76.

33. Ibid., p. 93.

34. Ibid., p. 95.

35. *The Evolution of an Evolutionist*, p. vi.

36. To the question "what evolves?" K. Boulding answers "something very much like knowledge"; *op. cit.*, p. 33.

37. *Synergism*, p. 49.

38. See Yves Barel, *La reproduction sociale* (Paris: Anthropos, 1973), pp. 63–127.

39. Ibid., p. 69 (my translation).

40. Ibid., p. 70 (my translation).

41. Ibid., pp. 100–102.

42. *Autopoiesis* (1980), p. 96.

43. *La méthode*, p. 179.

44. Y. Barel, *La reproduction*, p. 125 (my translation).

45. Ibid., pp. 125–126.

46. See R.E. Leakey and R. Lewin, *Origins* (New York: E.P. Dutton, 1977), and R. Leakey, *The Making of Mankind* (New York: E.P. Dutton, 1981).

47. S.M. Hines, "Biopolitics and the Evolution of Inquiry in Political Science," *Politics and the Life Sciences*, Vol. 1, No. 1 (1982), p. 6.

48. Anthony Giddens claims that "the notion of adaptation is so important in evolutionary theories that without it they lose most of their cogency." *The Constitution of Society: Outline of the Theory of Structuration* (Cambridge: Polity Press, 1984), p. 233.

49. The argument presented hereafter is inspired from, although it is not a direct rendering of, H. von Foerster, "Note pour une épistémologie des objets vivants," in E. Morin and M. Piatelli-Palmarini (eds.), *L'Unité de l'homme* (Paris: Seuil, 1974), pp. 401–417.

50. Ibid., p. 408.

51. Here I take issue with Habermas, who attacks Waddington for assuming that health is an objective value: "the biologist is in no way forced to adopt as his own preference the observed tendency to self-maintenance inherent in organic life—unless it be through the fact that he is himself a living being. But in the objectivating attitude of the knowing subject he can ignore this fact." Where could such an inhuman "objectivating attitude" lead but to a nihilistic view of the world as a lifeless emptiness inhabited by self-destructive intellects? Habermas, "Toward a Reconstruction of Historical Materialism" in *Communication and the Evolution of Society* (Boston: Beacon Press, 1979, translated by McCarthy), p. 177.

52. Edward O. Wilson, *Sociobiology: The New Synthesis* (Cambridge, Mass.: Belknap Press of Harvard University Press, 1975), p. 4.

53. See Richard Dawkins, *The Selfish Gene* (New York: Oxford University Press, 1976); R.L. Trivers, *Social Evolution* (Menlo Park, Calif.: The Benjamin/Cummings Publishing Co., 1985); E.O. Wilson, *Sociobiology*, and *On Human Nature* (Cambridge, Mass.: Harvard University Press, 1978).

54. See also Joseph W. Smith, *Reductionism and Cultural Being* (The Hague: Martinus Nijhoff Publishers, 1984), especially Chapters 4 and 5.

55. See Marshall Sahlins, *The Use and Abuse of Biology: An Anthropological Critique of Sociobiology* (Ann Arbor: University of Michigan Press, 1976).

56. See Steven Rose *et al.*, *Not in Our Genes* (New York: Penguin Books, 1984).

57. C.J. Lumsden and E.O. Wilson, *Promethean Fire: Reflections on the Origin of Fire* (Cambridge, Mass.: Harvard University Press, 1983), p. 121; see also C.J. Lumsden and E.O. Wilson, *Genes, Mind, and Culture: The Coevolutionary Process* (Cambridge, Mass.: Harvard University Press, 1981).

58. *Not in Our Genes*, p. 281.

59. On this point see Eugene F. Miller, "Positivism, Historicism and Political Inquiry," in *American Political Science Review*, Vol. 66, No. 4 (1972), pp. 796–797.

60. See F. Hayek, *The Road to Serfdom* (Chicago: The University of Chicago Press, 1944), and *The Counter-Revolution of Science* (Glencoe, Ill., 1952); see K. Popper, *The Poverty of Historicism* (London: Routledge and Kegan Paul, 1957).

61. See "Positivism," p. 800.

62. See *Ideology and Utopia* (London: Routledge and Kegan Paul, 1936).

63. On this see Pierre Birnbaum, *La fin du politique* (Paris: Seuil, 1975), pp. 71–77.

64. See G.H. von Wright, *Explanation and Understanding* (London: Routledge and Kegan Paul, 1971), pp. 1–3.

65. See McCarthy, *The Critical Theory of Jürgen Habermas* (Cambridge, Mass.: The MIT Press, 1979), pp. 169–170.

66. See Charles Taylor, "Interpretation and the Sciences of Man," *The Review of Metaphysics*, Vol. 25, No. 1 (1971); and M.J. Shapiro, *Language and Political Understanding* (New Haven: Yale University Press, 1981).

67. See A. Giddens, *New Rules of Sociological Method* (New York: Basic Books, 1976), p. 55.

68. On H.-G. Gadamer's work see ibid., pp. 55–65, and McCarthy, *Critical Theory*, pp. 170–183.

69. Gadamer, *Truth and Method*, quoted in McCarthy, *Critical Theory*, p. 175; a corollary of this position, to which McCarthy also calls our attention because it has been picked up by Habermas, is that hermeneutics has a practical dimension: "it is linked with transcendental necessity to the articulation of an action-orienting self-understanding" (quoted from Habermas' *Zur Logik der Sozialwissenschaften* in ibid., p. 179).

70. *New Rules*, p. 58.

71. See *The Social System* (New York: Free Press, 1951).

72. See T. Parsons, "Evolutionary Universals in Society," in *American Sociological Review*, vol. 29 (1964), pp. 339–357, and *Societies: Evolutionary and Comparative Perspectives* (Englewood Cliffs, N.J.: Prentice-Hall, 1966).

73. See in particular R. Nisbet, *Social Change and History* (New York: Oxford University Press, 1969), pp. 262–267.

74. McCarthy, *Critical Theory*, p. 234.

75. Ibid., p. 219.

76. See his "Historical Materialism and the Development of Normative Structures" and "Toward a Reconstruction of Historical Materialism," in *Communication and the Evolution of Society*.

77. Habermas, *Legitimation Crisis*, p. 7, quoted in McCarthy, *Critical Theory*, p. 234.

78. On this point, see Habermas, "Toward a Reconstruction of Historical Materialism" in *op. cit.*, p. 140, and McCarthy, *op. cit.*, p. 260.

79. In his review of Thomas McCarthy, *The Critical Theory of Jürgen Habermas*, *American Political Science Review*, Vol. 73, No. 3 (1979), p. 861.

80. For a witty critique of the idea of progress, see Geoffrey Vickers, *Freedom in a Rocking Boat* (Penguin Books, 1972), Chapter 1.

81. See J. Piaget, *Six études de psychologie* (Paris: Gonthier), pp. 174ff.

82. See "Historical Materialism . . . ," p. xxx.

83. See McCarthy, *Critical Theory*, pp. 250–253, and Habermas "Toward a Reconstruction . . . ," p. 156.

84. "Historical Materialism . . . ," p. 105.

85. "Towards a Reconstruction . . . ," p. 158.

86. "Towards a Reconstruction . . . ," p. 160.

87. This point is made by Floyd W. Matson, see *The Broken Image*, pp. 36–45.

88. Ibid., p. 41.

89. Ibid., p. 40.

90. On the primacy of political theory in Spencer's work, see David Wiltshire, *The Social and Political Thought of Herbert Spencer* (Oxford: Oxford University Press, 1978), p. 1.

91. For a sympathetic review of Spencer's evolutionary theory, see R. Puligandla, "The Concepts of Evolution and Revolution," in R. Gotesky and E. Lazzlo (eds.), *Evolution-Revolution* (New York: Gordon and Breach, 1971), pp. 41–70.

92. D. Wiltshire, *Herbert Spencer*, p. 197 (my emphasis).

93. Herbert Spencer, *On Social Evolution*, J.D.Y. Peel (ed.) (Chicago: University of Chicago Press, 1972), p. 47.

94. D. Wiltshire, *Herbert Spencer*, p. 197.

95. This point is made by F. Matson in *The Broken Image*, p. 40.

96. See J.D.Y. Peel in H. Spencer, *Social Evolution*, pp. xxviii–xxix.

97. T. Parsons, *Societies*, p. 21.

98. A. Giddens, *The Constitution*, p. 271.

99. As far as social evolution is concerned, these "universals" are i) the kinship system, ii) cultural legitimation, iii) bureaucratic organization, iv) money and markets, v) a universalistic legal system, and vi) democratic associations.

100. A similar distinction between "specific" and "general" evolution is made by M.D. Sahlins and E.R. Service in *Evolution and Culture* (Ann Arbor: University of Michigan Press, 1960); T.L. Thorson attaches a great deal of importance to this distinction, see *Biopolitics* (New York: Holt, Rinehart, and Winston, 1970), pp. 140ff.

101. Namely, bureaucratic organization, money and markets, a universalistic legal system, and the democratic association in both governmental and private forms.

102. See McCarthy, *Critical Theory*, pp. 181–193.

103. On this point, see "Toward a Reconstruction . . . ," p. 130.

104. Habermas refers twice to Morin's German translation of *Le paradigme perdu: la nature humaine* (Paris: Seuil, 1973).

105. "Toward a Reconstruction . . . ," p. 135.

106. Ibid., p. 137; for a development of a similar idea, see E. Morin, *Le paradigme perdu*, pp. 75–91, and C.L. Brace, "Biological Parameters and Pleistocene Hominid Life-Ways," in I.S. Bennstein and E.O. Smith (eds.), *Primate Ecology and Human Origins: Ecological Influences on Social Organization* (New York: Garland, 1979); however, on the basis of anatomical evidence, other authors claim that language is actually a more recent development.

107. "Toward a Reconstruction . . . ," p. 134.

108. See E. Morin, *Le paradigme perdu*, pp. 171–180.

109. "Toward a Reconstruction . . . ," p. 138.

110. Ibid., p. 147.

111. Ibid., p. 148.

112. Moreover, Habermas insists that the "base" does not coincide with the economic system, except in capitalist societies (see ibid., p. 144).

113. Ibid., p. 148 (my emphasis).

114. H. Frankfort, *The Birth of Civilization in the Near East*, p. 119, quoted by E.R. Service in *Origins of the State and Civilization: The Process of Cultural Evolution* (New York: Norton, 1975), p. 231.

115. E. Morin, "Le complexe d'Adam et l'Adam complexe" in *L'Unité de l'homme*, p. 743 (my translation).

116. See Y. Barel, *La reproduction*, p. 126.

117. Ibid., pp. 409–413. Boulding also devotes much attention to the production and reproduction of artifacts—too much attention in fact. To draw an analogy between the automobile and a biological species and to describe the blueprints designed by engineers as the genotype of that artificial species could have been the starting point of a reflection on the fundamental difference between the allopoietic character of artifacts as opposed to the autopoietic character of life and of the "noosphere." On this subject (of the noosphere), he proposes some interesting but impressionistic ideas about man's ability to make (and to make evolve) images of his world as well as of other worlds, and thus to project himself into a preconstructed future. Unfortunately, these insightful points remain unrelated in the absence of a coherent theoretical framework; see *Ecodynamics*, pp. 121–140.

118. *La reproduction*, pp. 413–420.

119. Ibid., pp. 420–421.

120. Ibid., pp. 421–424.

121. *Synergism*, p. 28.

122. Corning's use, if not abuse, of the concept of synergism has been sharply criticized by C.R. Hess in his review of *The Synergism Hypothesis* (*Politics and the Life Sciences*, August 1984, pp. 102–105; see also Corning's reply in the February 1985 issue of the same journal).

123. Ibid., p. 236.

124. See E. Morin, *La Méthode*, vol. 2: *La vie de la vie* (Paris: Seuil, 1980).

125. Ibid., p. 66 and p. 67 (my translation). (Biocoenosis is defined by Webster's Dictionary as "a community of biologically integrated and interdependent plants and animals.")

126. Ibid., p. 237. In an earlier work, Morin had attempted to show that the human nature emerges at the junction between a self-organizing biosphere realizing its invariant projects (e.g., the genetic codes of the various species) by reintegrating their deviation (e.g., mutations) and a man-made world where historical projects seemingly evolve towards universality (e.g., the scientific culture) through differentiation, competition, and conflicts. See *Le paradigme perdu: La nature humaine* (Paris: Seuil, 1973).

127. The inescapable dominance of language over the scientific pursuit of knowledge was a theme investigated by Arthur F. Bentley, long before it became a fashionable topic in the philosophy of science; see J.F. Ward, *Language, Form and Inquiry: Arthur F. Bentley's Philosophy of Social Science* (Amherst: The University of Massachusetts Press, 1984).

5

Scope and Methods of an Organizationalist Approach to Political Inquiry

The political universe rarely unfolds as it should. Neither the theory of self-organizing systems, nor the epigenetic paradigm of biological and/or societal evolution, are readily applicable to the study of concrete political problems. But political scientists should be able to derive from these research programs new methods of inquiry into the immense complexity of political life. They may even learn from the practice of such methods to see their whole subject matter in a different light; one that, far from being radically new, reveals neglected or unexpected connections between empirical and normative theory.

In all things methodological, the principle of "monster conservation" should apply, to use M. Thompson's amusing phrase.[1] Two such monsters will be ferreted out hereafter, but saved from the big guns of the conventional "scientific method": time and dialectic. Much of this discussion revolves around the articulation of a method that would permit a non-reductionist, non-dogmatic approach to logical paradoxes and to the paradoxes of life in particular. The last section proposes a vision of political life as an evolutionary process and outlines a three-dimensional analytical grid for interpreting the paradoxes that such a process entails. More specifically, the reference to the three dimensions of consciousness is meant also to facilitate the exploration of possible connections between the paradigmatic notions of system and "life world." (The latter is central to those social theories that insist on the primary importance of treating social agents as free thinking subjects—a view that is not incompatible with the kind of non-mechanistic systems paradigm developed in the previous chapters.) The next three chapters flesh out this analytical framework.

Toward a Diachronic Method

Anthony Giddens claims that there is no room for synchronic models in social theory: they are inherently faulty.[2] His criticism is particularly directed at functionalism and structuralism which he attacks for their reductionist identification of change with the unstable and of stability with timelessness. He is most convincing when he argues that the artificial opposition of the static (i.e., synchronic) and dynamic (i.e., diachronic) modes of analysis makes it very difficult, if not impossible, to pose the problem of the continuity of the field of historicity. Contrary to Parsons, who overestimated the contribution of motivated acts to the furthering of societal change, Giddens envisages traditional, routinized cognitive processes as the medium of societal change. A significant transformation takes place when reproductive processes become thematized as problematical or, as he puts it, when "de-routinisation" occurs.

When Giddens writes, "the conception of structuration introduces temporality as integral to social theory"[3] he might as well have replaced "structuration" by self-organization. What follows from this premise regarding, as we have just seen, the dynamic character of supposedly static phenomena and the consequent need for a generalized diachronic approach applies equally to my neo-cybernetic paradigm. But I would like to take issue with Giddens' overextended formulation of his basically sound premise: temporality loses its significance if one cannot retain the analytical tension which it presupposes. To deny, under any circumstance, the validity of a synchronic approach amounts to saying that there never arises a need for a critical reflection on the actual manifestation of temporality. Thus we are back to square one, so to speak. Besides, it is impractical.

No research project on the self-organization of political life is complete until specific hypotheses about developmental paths and the corresponding autopoietic mechanisms have been specified. However, the relative importance of these hypotheses in the overall research agenda may vary, provided that justifications are offered for the degree of emphasis that the researcher decides to attribute to either the short-term analysis of structural relations or to the diachronic exploration of organizational potentialities.

Thus, socialization cannot be treated as a specialized area of research but must be conceptualized as a crucial dimension in *all* studies of cultural patterns. Empirical surveys should always be designed so as to provide us with at least some data about the extent to which normative orientations are shared with other members of the political community: the frequency with which an opinion is voiced, to whom, under what circumstances, etc., or identification by the respondent of the sources of his information, valuations, etc. One could also try to look for the emergence of new routines involving the critical re-evaluation of implicit norms (e.g., changes in vocabulary: from "Negroes" to "Blacks," from "*Canadien*" to "*Québécois*," etc.).

Should political theory become more concerned with the self-organization of political life, that is with the historicity of politics, then the question of its relation with history (as a discipline) would arise. While I agree with Giddens' categorical assertion that "there simply are no logical or even methodological distinctions between the social sciences and history—appropriately conceived,"[4] it clearly needs to be qualified. History is primarily concerned with singular events, even if it has moved slowly towards the investigation of the political, economic, demographic, and even climatological patterns underlying historical contingencies. By contrast, even in its pre-behavioral days, political science has always aimed at the discovery of generalizable observations (e.g., what is the "true" nature of the state). Can these two visions be reconciled? Here again the epistemological loop is instructive; we are at present witnessing what Morin calls "the return of the event" in the natural sciences:[5] singular, unpredictable random occurrences are no longer excluded from the conceptual world of physics and biology—indeed they have come to play an increasingly important role in it (e.g., the interest in the concept of "chaos" in theoretical physics). Therefore, political scientists should not fear the danger of contamination by history; the scientificity of their methods does not necessarily depend on their predictive power. On the contrary, it is the dogmatic synchronism characteristic of most empirical models of political analysis which threatens their scientific relevance. Practically speaking, this means that political scientists should devote more attention to singular events and especially to political crises. Not that there exists no literature on events like the May 1968 student revolt in France, the proclamation of the War Measures Act in Quebec in October 1970, or the Watergate affair, but more often than not it has been written by journalists. In particular, we need to focus on the impact that those crises might have and on i) the regulatory mechanisms of the state (e.g., establishment of new bureaucratic agencies, adoption of new legislation, implementation of new law enforcement techniques, etc.), ii) the legitimacy of state institutions.

Dialectics Without a Synthesis

A social scientific theory of self-organization is intrinsically paradoxical. A paradox consists of a message that bears upon itself. From a more formal viewpoint, one cannot define a set which is also an element of itself without proposing a paradox (e.g., Russell's paradox about the barber who shaves all those who do not shave themselves: who then shaves the barber?).[6] A self-organizing system is paradoxical since it exists as an object and a subject like a mythical snake eating its tail. The reproduction of the same by the same, in biological as well as in political life, implies that the form that was senses in itself the form to be, as if it were able to read its own message from an outside that would be located inside. All theories that are concerned with recursive phenomena are based on a paradox to which Y. Barel has given the generic name "*le paradoxe de l'auto . . .*"[7] Note that, in Anthony Giddens' opinion, the most relevant aspect of the new cybernetics (viz. Varela's work on autopoiesis) is precisely that it is a method for conceptualizing recursive phenomena.[8] This author intuits that there may exist an interesting connection between biocybernetics and the social sciences which remains to be examined. He adds that "however this may be in biological systems, I shall want to argue . . . that the self-regulating properties of social systems must be grasped via a theory of *System Contradiction*."[9] But I claim that he should have written instead a "theory of system paradoxes"; and, if so, we first have to learn more about paradoxes and how to use them in the logic of inquiry before we can even begin to construct such a theory.

Confronted with a paradox, philosophers adopt either one of two possible attitudes. They may try to show that the paradox is not really a paradox, that it is, after all, possible to distinguish unequivocally between the message and the meta-message containing the rules of interpretation. Thus Plato, according to A. Glucksmann,[10] tried to prove that Socrates could have said meaningfully that he knew that he knew nothing because, having once left the cave for the luminous space that lies *above*, he returned with a meta-knowledge that permitted him to dispute the shadowy opinions we mistake for true knowledge. The "master thinkers" are those who claim to possess such a superior knowledge, those who dispell the paradoxes, "solve" the contradictions.

A minority of theorists accept the irreducibility of paradoxes and thus do not succumb to the temptation of building total and inevitably totalitarian systems. This is the strategy that I consider to be the most consistent with the self-organization paradigm insofar as the latter attempts to be a cybernetics of cybernetics, but does not purport to recover at the meta-level the positivistic assumptions for which it faults the first generation of cybernetic models.

The Paradox of Life

Uncertainty, ambiguity, contradiction, paradox: these terms are generally used indiscriminately and often inappropriately. Yves Barel, whose ideas are discussed *infra*, has done much to clarify the notions of contradiction and paradox.[11] As stated above, a paradox entails a tension between two distinct levels of discourse. A contradiction, on the other hand, develops on a single level. It takes the form of an opposition between antithetic but partially overlapping terms (e.g., Vickers' "doers" and "done-by," Roberto Michels' "iron law of oligarchy," etc.). To the extent that a contradiction tells us less about the structure of the space of representations than a paradox, as it were, its analytical usefulness is more limited than that of a paradox. Moreover, a contradiction constitutes a static relationship since, by definition it is not possible to transcend it by moving onto another level in the space of representations. In other words, a contradiction can never be "solved"; this is a problem that plagues Marxist theory for, according to Anthony Wilden, the capitalist system could not possibly collapse if its internal contradictions were nothing else than that.[12] If it is not so, it is because Marx's "contradictions" really operate as paradoxes; as Barel puts it: "It is, as it were, in trying to be what it is not, in order to be what it is, that capitalism destroys itself."[13] However, Wilden's emphasis about the necessity to distinguish among the various levels of discourse almost leads him to take away from paradoxes their paradoxical character. Barel argues that, while all contradictions do not qualify as paradoxes, all paradoxes are nevertheless contradictory. There always remains a point where the two levels merge and thus where opposite terms may enter into contradictory relations. Moreover, there is no reason to posit that, in any given contradiction, the terms are themselves simple; they may subsume other contradictions or even paradoxes (e.g., illness and cure in I. Illich's *Medical Nemesis*[14]).

It is vitally important (that is to say, it is a matter of life and death for the social sciences *qua* sciences) to realize that in many fields and in many ways, research programs are being developed in order to deal with paradoxes as the proper subject matter of scientific inquiry and not as mere logical aberrations. Although these methods are still often tentative (e.g., the new cybernetics), if not altogether unorthodox, and have not yet been articulated in a manner that is directly relevant to the concerns of social scientists (with the possible exception of E. Morin's recent works), they already offer some precious analytical instruments (see the next sub-section). Together they form a "dialogical" methodology, to borrow a phrase from von Foerster. Such an approach is probabilistic and dialectical—that is, if we do not equate dialectic with the Hegelian-Marxist method. Indeed a dialogical cognitive strategy, if it is used imaginatively, could take us beyond the strictures of historical materialism.

These strictures are both metaphysical and methodological. From a metaphysical point of view, Marx and Engels did not so much put Hegel's dialectical method back on its feet as altogether deprive it of any ontological foundation.[15] First, there is the problem of the "materialistic" axiom: it has become trivial to stress that matter is nothing like what 19th century positivists thought it was. But it may not be reasonable to criticize a theorist for his lack of prescience. More importantly, Marx and especially Engels, who is here the principal offender by making dialectic an attribute of nature itself rather than a characteristic of discourse,[16] reduced Hegel's dialectic to a tautology: dialectic is a scientific discourse because it is a scientific discourse about dialectic![17]

Furthermore, dialectic is itself reduced to three "general laws of motion."[18] This nomological *coup de force* explains everything—and thus explains nothing because no logical system can be self-inclusive unless it is ultimately actuated by the tacit (i.e., illogical) knowledge of knowing subjects in a manner which cannot be expressed by any universal law.

By contrast, the emergent cosmology sketched out by Morin on the basis of his encyclopedic survey of contemporary developments in thermodynamics, biology, cybernetics, etc., uses something like a dialectical language for expressing the incommensurable, recursive, and redundant complexity of universal processes. However, it does not do so in order to codify nature once and for all. For Morin, dialectics, i.e., the deliberate use of opposite terms placed in interactive opposition, sustains the dialogue that we enter into, as it were, with the autopoietic "beings" surrounding us, of which we partake and upon which we act (e.g., the stars, the cells, animal societies, human beings, and human groups, etc.). Only by reifying these autopoietic "beings," that is to say, by taking away from them precisely that which makes it necessary to resort to a dialectical method, could we specify *a priori* the "laws" governing their evolution.

From a methodological point of view, Marxist dialectic consists in a juxtaposition of supposedly homogeneous elements (e.g., "capital," "labor," etc.) in such a way that the tension which is assumed to exist between them is "resolved" by the eventual and inevitable substitution of one by the other. But I concur with John W. Sutherland who writes, "the societal dialectic is most often an incomplete one: we find societal systems pursuing either thesis or antithesis, on virtually all dimensions, while instances of synthesis remain scarce indeed."[19] The method advocated here aims at identifying the contradictions within each contradiction, as it was suggested above;[20] Barel speaks in this respect of "the terms of the terms" of a contradiction.[21] In the analysis of a complex system, we ought to multiply the opportunities for articulating recursive and redundant connections (redundancy meaning here a situation in which a part entails the whole—without altogether entailing it!).

Yet, no doubt paradoxically, a neo-cybernetic approach to politics cannot but share some notions, methods, and concerns with historical materialism. First, as I have noted already, there is little point in rediscovering what has been known for long. Marxist scholarship offers us a wealth of theoretical and/or empirical research on some crucial conflicts in industrial societies, between developed and underdeveloped regions, etc.

Second, and more importantly, historical materialism has already delineated some of the most crucial problems that an evolutionary perspective must focus upon (e.g., the reproduction of the forces of production, the reproduction of the social structures of domination and of their legitimizing discourses, etc.). Indeed the numerous references to Marx himself or to his more or less critical, more or less revisionist followers in the previous chapters are proof enough of this convergence.

Yet it remains that historical materialism is an ideology—some would even say *the* archetypical ideology: a secular eschatology. The organizationalist paradigm is not. Wherever the former introduces a closure (e.g., the "resolution" of contradictions, the "laws" of dialectics, the primacy of the "infrastructures"), the latter will strive to force it open. The organizationalist paradigm challenges the analyst into confronting complexity without first reducing it to something else, for it does not provide any answer; it is strictly a method of inquiry.

The Logic of Autonomy

The preceding methodological prescriptions are likely to produce a profusion of alternative interpretations and contradictory data. It behoves the observer to straighten things out by indicating what ordering pattern is preferrable and working out its implications. Order is not arbitrary (at least not essentially so); nevertheless, it is created subjectively and validated discursively in a community sharing a certain historical and philosophical tradition, a language, etc. Thus I adhere without reservation to A. Giddens' following "rule" (where we may substitute "political" for "sociological"):

> The sociological observer cannot make social life available as a "phenomenon" for observation independently of drawing upon his knowledge of it as a resource whereby he constitutes it as "topic for investigation."[22]

As Giddens remarks, "knowledge" in the present context does not refer to the storage of discrete "corrigible items" but to "interpretative schemes." This I take to mean that the way in which the observer orders his values is reflected in his interpretation of political life. Even though the inherently value laden character of political inquiry needs to be reaffirmed, it is not an original idea. What is new is Giddens' position (although he is not alone in defending it) is that he does not consider the observer's values as an unwelcomed intrusion upon his/her work. On the contrary, he regards them as prerequisites for achieving "the hermeneutic explication and mediation of divergent forms of life within descriptive metalangues of social science."[23] Values attest the humanness of both the societal actors and the social scientists.

Therefore, a scientific theory ought to be formulated in a manner that enables its users to *reconstruct* the interpretative scheme from which it is derived. This entails more than the ritualistic caveat to the effect that one should state one's values explicitly. (It seems to be usually taken for granted that such a confession suffices to exorcise the scientist of his/her subjectivity and thus to enable him/her to carry out his/her observations objectively!) The observer's values should be formalized to the greatest possible extent in order to facilitate their articulation to the body of the theory.[24] Depending on how far one wishes to retrace one's analytical steps, various techniques may be used separately or in conjunction. I shall be concerned hereafter more specifically with the formalization of the logical foundations of theoretical research, using the logical underpinning of this book as an example. The problems posed by the application of a social scientific theory to policy analysis are briefly discussed in the appendix where I argue that some techniques of computer simulation make it possible to work out and critically examine the implications of the various ways in which paradoxes can be "solved" in order to initiate and implement political decisions.

Ordinarily, theorists do not specify the logical rules that they have followed in their work. It is simply assumed that the traditional canons of the two-valued, Aristotelian logic obtain. But this implicit assumption needs to be justified, for Aristotelian logic is no longer the only logic that scientists may use; indeed its usefulness appears to be severely limited.

For A.J. Bahm, the paradoxical nature of complex systems necessarily implies that systems theory must turn away from traditional logic if it is to remain true to its holistic premises.[25] This is all the more important to stress because systems analysis, i.e., the kind of decisional technology which more or less directly echoes the concepts if not the assumptions of systems theory, presupposes the universal applicability of disjunctive operations in problem solving. However, G. Günther argues that cybernetic systems behave as subjects, not because of some metaphysical essence, but insofar as their operations entail "the logical structures of transjunction";[26] this phrase refers to the ability on the part of cybernetic systems to reject both terms of an alternative "objectively" describing reality (e.g., the conjunction of two events is necessary or not necessary).

Therefore, social scientists need to mobilize the still somewhat meager but growing conceptual resources of multi-valued logics.[27] As early as the 1930s, J. Lukasiewicz proposed a logical mode of analysis that permitted the use of truth-values other than true and false. Of course, one could look even further back to Hegel, Fichte, or Schelling whose reflective, dialectical approaches to philosophy in general, and to logic in particular, prepared the epistomelogical ground for modern non-Aristotelian logics. Unfortunately, their systems were insufficiently formalized. The need for such methods has not been felt in science until recently—that is until the articulation of the research program of the new cybernetics. Notwithstanding G. Günther's impressive contribution to "trans-classic" logic, to use his own term, I now turn to F. Varela's formulation of autopoiesis, for I consider it to be more directly relevant to my present concerns.

Varela has worked out a three-valued logic on the basis of George Spencer Brown's algebra.[28] The truth-values are as follows: Brown's "marked" and "unmarked" states, respectively symbolized by the operators "⌉" and " " (blank), and his own concept of autonomy to which he associates the operator "□". (For Brown the operation of demarcation refers to the *act* by which an observer builds up his world-view;[29] Varela's "□" indicates a self-referential demarcation.)

Political scientists may find it difficult, if not impossible, to apply Varela's calculus to their fields of inquiry. However, even at a verbal level, this calculus can be useful in that it allows one to conceptualize autonomy as distinct phenomenality. For example, my choice of the three levels at which the creation and destruction of order has been analyzed can be better understood by reference to Varela's truth-values: experience entails the demarcation of systems from their environmental background (i.e., using the operators " " and "⌉"); consciousness arises as a result of the demarcation between the objective and subjective dimensions of one's experience (i.e., using the operators "⌉" and "□"); eventually, self-consciousness takes the form of a critical re-evaluation of the relations established at the two lower levels and thus develops in a three dimensional space (i.e., using the three operators together).

The Evolutionary Dynamics of Political Life

The Multidimensional Logic of Political Inquiry

Definitions of politics abound and I do not intend to propose yet another one here, at least not an unequivocal specification of the concrete or abstract boundaries of something "out there" called politics. Because politics is typically complex, it can only have fuzzy boundaries. My concern is with the subject-matter of a science of politics and my purpose is to show that it encompasses all modes of societal interactions. This is not because everything that is social is *ipso facto* political, but rather because politics is central to the process of societal reproduction.

Empirical political theory is predicated on the misguided assumption that political behavior is easily distinguishable from other types of social behavior. Political life is defined as the particular function of a specific societal system interacting with other such systems. And political science, economics, sociology, etc., are conceptualized as distinct scientific disciplines. Of course, this has not always been so; let us recall that for Plato politics, philosophy, and what we would now call the natural sciences were homological, if not coinciding, domains of inquiry. A commonly adduced reason for the segmentation of scientific knowledge is the need to specialize research. Indeed, no one could become an expert in all the subfields of any contemporary science, let alone in all the scientific disciplines. However, the practical and epistemological consequences of this kaleidoscopic view of nature and of society are very serious. In particular, the reduction of politics to "political behavior" results in the aberrant application of sophisticated research techniques to the investigation of trivial questions. Admittedly, this situation has changed somewhat over the last decade; nowadays it looks almost as if "anything goes." In the wake of the "post-behavioral revolution," political scientists have adopted a wide variety of epistemological positions (e.g., from phenomenology to neo-Marxist political economy), of methodological orientations (e.g., biopolitics, political linguistic, etc.), and of empirical concerns (e.g., the exponential growth of public policy studies). But too often these new directions have been presented as *the* road to salvation for political science, or have resulted into a further segmentation of the discipline (especially in the field of policy studies).

This new eclecticism is a healthy sign of renewal; one could perhaps see there the "noise" required to throw political science on a new course. But now is the time to channel this new energy. If everything is interesting, nothing is worth studying in depth; if everything is relevant, nothing can be significant. The temptation is to make arbitrary pronouncements about *the* new goal of political inquiry. It seems more promising to search for new ways of looking at the origins and nature of political order. To some this may sound like a belated, if not naive, rediscovery of the Hobbesian problem but I shall argue that this is not the case. Stated as such, however, this problem alludes to the eventuality of opening a much needed dialogue between political philosophers and empirical theorists, although hereafter a greater emphasis will be placed on the discussion of issues relevant to the concerns of the latter.

The behavioralist movement cannot be faulted for ignoring the problem of order. Indeed democratic theory contains some interesting paradoxes. For example, G. Almond and S. Verba argue that participatory democracy is possible only on the condition that most political actors participate only infrequently to politics![30] But the problem is inadequately posed. It is assumed that political actors know the rules of the game, and therefore one is left with the problem of explaining their behavior within what appears as a given context. However, the question of knowing where these contextuating rules come from, how they have evolved, and how they can be changed is hardly raised.[31] My contention is that this should be a dominant concern for political scientists.

The first step in this direction consists of rethinking the "level of analysis" problem. It is usually presented as a dichotomous choice between micro-analysis—at the level of individuals or groups—and macro-analysis—at the

LEVEL OF ANALYSIS	PROCESSES	STRUCTURES	FORMS OF ANALYSIS
Experience	Regulation (allocation of values; goal achievement)	Political System	Cybernetics
Consciousness	Autopoiesis (Selection and ordering of collective norms)	Polity (nation-state)	Theory of self-organizing systems
Self-Consciousness	Evolution (critique of political theory)	Civilization (tradition)	Theory of knowledge (cybernetics of cybernetics)

Table 5.1

level of the political system.[32] But macro-analysis involves more than looking at the performance of political institutions in their supposedly given societal context; it also involves an investigation of the evolutionary political dynamics that molded these institutions. From what has been established already, it should be clear that the theory of self-organizing systems offers useful suggestions in this regard.

Political order is an organizational capacity that operates through a succession of destructuring and restructuring phases; it can be conceptualized as a multi-level process as Table 5.1 illustrates.

The rationale for the distinction made between the first two lines in the above table is best understood with reference to the notions of *Gemeinschaft* and *Gesellschaft* (which are introduced immediately hereafter). These methodological questions are further discussed in the next two chapters. The last level of analysis (i.e., self-consciousness), whose inclusion here reflects a concern for the place of political theory itself in the evolutionary process, raises a wealth of epistemological and even ontological problems that could not possibly be treated exhaustively within the limits of this book. Nevertheless, I attempt in Chapter 8 to sketch out a naturalized epistemology that would facilitate the elaboration of a critique of man's exploitative orientation towards both nature and himself.

On the Nature of the Body Politic

We owe to Ferdinand Toennies the classical distinction between the highly differentiated *Gesellschaft*, i.e., "a limitless totality of a system of conventional rules,"[33] brought about by industrialization, and the *Gemeinschaft* consisting of "lasting and genuine forms of living together."[34] Toennies "contrasted the idyllic *Gemeinschaft* [of a by-gone preindustrial age] with the repugnant mercantile, conflict-ridden Gesellschaft."[35] Stripped of their ideological overtones, these notions may still be instructive. Contemporary social theory focuses almost exclusively on society and ignores communities, when actually both dimensions may coexist. Such a methodological assumption deserves to be given more consideration in view of the parallel that Morin establishes between the *Gesellschaft* and negative feedback control models of society and, on the other hand, between the *Gemeinschaft* and self-organizing systems.[36]

If, after the example of the evolutionists, we wish to "take time seriously,"[37] we ought to pay equal attention to the short and long term aspects of political life. First, we can think of politics as the outcome of the political system, defined as a differentiated steering subsystem within a given *Gesellschaft* (see Chapter 6). Second, we can take a more comprehensive point of view and look at the same human community as constituting a self-organizing, undifferentiated *Gemeinschaft* emerging out of communicative interactions, which are often of a political nature but are not confined within a clearly demarcated political system. Neither of these perspectives is entirely satisfactory if adopting one means rejecting the other.

Either the political system is the locus of politics, but it is not because politics is an autopoietic process rather than the attribute of a structure; or it is not, and then it is a system without substance. Let us recall H. Alan's characterization of a self-organizing system as being highly redundant. For example, the neocortex of the human brain can reorganize itself after having been partially destroyed. The rigidly structured *Gesellschaft* consisting of institutionalized subsystems, including the political system, does not conform to this standard. In other words, the evolution and inner-logic of political phenomena cannot be understood by reference to a political subsystem isolated from its societal matrix. At the same time, the complexity of societal relations which is the condition of political change is a corollary of the differentiation of societal norms and of their administration by specialized subsystems (e.g., the political system).

There is no reason to regard the *Gesellschaft/Gemeinschaft* dichotomy as anything else than a preliminary approximation of complex and therefore elusive notions. Indeed, Max Weber's reformulation of this dichotomy in dynamic terms (i.e., *Vergesellschaftung/Vergemeinschaftung*) appears to be more consistent with the approach adopted here.[38] On the other hand, Emile Durkheim's understanding of the same problem in terms of two types of solidarity, i.e., "mechanic" vs. "organic," corresponds even more closely to my premises. While Toennies thought that the organic unity of communal life had been lost with the transition to modernity, Durkheim argued that such a "solidarity," as he puts it, develops as a consequence of modernization.[39]

As far as the political dimension of a society's organic solidarity is concerned, Morin claims that a nation is a *living* reality, since it consists of connections linking the neurocerebral systems of its individual members; it is also, and above all, a collective psychic reality, even a "psychically anthropomorphic being."[40] (In Morin's view, political structures such as the state, or the "party-state" in non-liberal democracies, form the nucleus or "genostructure" of this being.) This analogy suggests two distinct, although interrelated, directions of analysis. The first would take the form of an inquiry into the reproduction of national political cultures, and of the appropriation of cultural values by the state (it is taken up in Chapter 7). The second would focus on the reproduction of a "biospheric" body politic, conceived as a biological and socio-technical complex depending on a larger environment and whose organization has now become a political problem in its own right, as we cannot go on thinking that nature, in its infinite bounty, exists only to be pilfered or domesticated. Some of these questions and their implication for political theory are addressed in Chapter 8.

Morin's approach is suggestive in many respects. First, it clearly posits that political life ought not to be limited to a specific domain but plays a central role in the production of a formative image of society as a whole. Second, it offers an opportunity to analyze the political implications of a paradigmatic vision of man as an image-maker, which seems to be gaining ground in anthropology, as the previous chapter has shown. It is also consistent with the emphasis that the new cybernetics places at whatever level, on recursiveness and reflectivity, from the methodology of model-building to logic. Morin's understanding of society as a self-organizing system with a political nucleus also bears a resemblance to Alain Touraine's view of society as the product of its own labor and of social relations structured by political decisions.[41] But Touraine examines the political process almost exclusively in terms of its regulatory practices. In contrast, Morin's emphasis on the noogenetic aspects of politics appear to be more original. Yet these can only be understood with reference to the direct experience that we have of political institutions as specialized instruments of societal regulation. Thus I turn in the next chapter to a discussion of the regulatory function of the state.

Notes

1. M. Thompson, *Rubbish Theory: The Creation and Destruction of Value* (Oxford: Oxford University Press, 1979), Chapter 7.
2. *Central Problems in Social Theory* (London: Macmillan, 1979), Chapter 6.
3. Ibid., p. 198.
4. Ibid., p. 230.

5. E. Morin, "Le retour de L'événement," *Communications*, vol. 18 (1972), pp. 6–20, the ultimate (and not quite orthodox) consequences of these new trends is the hypotheses derived from quantic theory according to which the causes and effects of an event, in the universe of the infinitely small, can be located anywhere in a "reversible time."

6. Actually, the barber's dilemma is a less profound, albeit more amusing, paradox than the one that it is meant to illustrate here. The former, proposed by Russell in 1918, is what Quine calls a "veridical paradox," i.e., one that "packs a surprise, but the surprise quickly dissipates itself if we ponder that proof." (W.V. Quine, *The Ways of Paradox*, Cambridge, Mass.: Harvard University Press, 1976, p. 9); there is indeed no place where such a barber practices his paradoxical talents. The latter, worked out by Russell in 1901, is a genuine antinomy, i.e., a proposition that "establishes that some tacit and trusted pattern of reasoning must be made explicit and henceforward be avoided or revised" (ibid., p. 5).

7. "Auto . . ." as in "*auto-réglage*" (self-regulation), or "*auto-organisation*" (self-organization), see Yves Barel, *Le paradoxe et le système* (Grenoble: Presses Universitaires de Grenoble, 1979).

8. Anthony Giddens, *Central Problems in Social Theory* (London: Macmillan, 1979), pp. 75–76.

9. Ibid., p. 76.

10. *The Master Thinkers* (New York: Harper & Row, 1979), Chapter 3.

11. *Le paradoxe*, Chapter 2 and *in passim*. Giddens, on the other hand, has done much to clarify the difference existing between contradiction and conflict, two notions that Marx (and most Marxist theorists) used as almost interchangeable terms; for Giddens, a conflict develops in the context of what he calls "social integration" (i.e., "systemness on the level of face-to-face interaction") whereas a contradiction occurs through "system integration" (i.e., "systemness on the level of relations between social systems or collectivities"); in other words, the former is an actual practice, the latter refers to a "disjunction of structural principles of system organization" (*Central Problems*, pp. 76–77, 131, and 144).

12. A. Wilden, *System and Structure* (London: Tavistock, 1972).

13. Ibid., p. 69.

14. I. Illich, *Medical Nemesis* (New York: Bantam Books, 1977).

15. What follows is derived from Alain Besançon, *Les origines intellectuelles du léninisme* (Paris: Calmann-Levy, 1977), pp. 51–56.

16. See *Anti-Duhring* and his posthumous *Dialectics of Nature*.

17. There is a fundamental difference between a tautology and what Morin calls a "loop." The latter is a dynamic relationship representing a recursive process, the former an *a priori* equivalence between identical categories. Morin's "loops" pose a real problem, tautologies solve in vain trivial problems.

18. These are: i) the law of the interpenetration of opposites; ii) the law of the negation of the negation; and iii) the law of the transformation of quantity into quality.

19. J.W. Sutherland, *Societal Systems: Methods, Modeling and Management* (New York: Elsevier North Holland, 1978), p. 84.

20. Morin writes "'*la pensée complexe*' accepts the challenge of contradictions. It could not consist, as it is the case with dialectics, in the 'transcendence' ('*Aufhebung*')

of contradictions. It consists in their desoccultation, in their display, and calls for wrestling with contradictions" (*La méthode: la vie de la vie*, vol. 2, Paris: Seuil, 1980, p. 285).

21. *Le paradoxe*, pp. 71–72.

22. A. Giddens, *New Rules of Sociological Method* (New York: Basic Books, 1976), p. 161.

23. Ibid., p. 162.

24. K. Deutsch has expressed a similar rule but at a still higher, if not unreachable, level of abstraction: the contexts of perception, relevance and interpretation "must be capable of being shared"; he continues: "They must be stateable, retraceable, and reproducible, by operations which can be performed by any one with requisite training. That is to say, they must be formally rational." ("On Political Theory and Political Action," *American Political Science Review*, vol. LXV, no. 1 (1971), p. 12).

25. See A.J. Bahm, "Systems Theory: Hocus Pocus or Holistic Science?," *General Systems Yearbook*, vol. XIV (1969).

26. G. Günther, "Cybernetic Ontology and Transjunctional Operations," in M.C. Yovits, et al., (eds.), *Self-Organizing Systems* (Washington, D.C.: Pergamon Press, 1962), p. 352.

27. G. Günther contrasts the postulates of classic (Aristotelian) and "trans-classic" (post-Aristotelian) logics in the following way:

classic	**trans-classic**
1) the dichotomy of form and matter is relevant for any system of pure logic;	1) the dichotomy of form and matter does not hold in n-valued systems when $n>2$;
2) the concept object is not ambiguous;	2) the concept $<$ object $>$ is amphibolic when $n>2$;
3) the semantic relation between Truth . . . Falsity (including a scale of intermediate terms) and logical value is unique.	3) the disjunction truth/falsity applies as value designation if and only if $n=2$.

"The Tradition of Logic and the Concept of Trans-classic Rationality," *Allgemenen Nederlands Tijdschrift voor Wisjsbegeerte en Psychologie*, vol. 54, no. 4 (1962), pp. 195–196.

28. See F. Varela, *Principles of Biological Autonomy* (New York: Elsevier, 1979), *in passim*.

29. "It may be helpful at this stage to realize that the primary form of mathematical communication is not description, but injunction"; G. Spencer Brown, *Laws of Form* (London: George Allen, 1969), p. 77.

30. See G. Almond and S. Verba, *The Civic Culture* (Boston: Little, Brown, 1965).

31. Of course, there is a sizeable literature on political development in which the transformation of political structures and traditions is approached from several theoretical perspectives. However, it is significant that a) the problem is not raised in the case of modern societies, and b) the literature on political development has become rather sterile lately (especially on comparative administration).

32. This is a simplification; Heinz Eulau, for instance, has argued that there are many intermediate levels and that the problem of understanding the multiple relations existing among these levels has not always been adequately posed. See H. Eulau, *Micro-Macro Political Analysis* (Chicago: Aldine, 1969).

33. S. Chodak, *Societal Development* (New York: Oxford University Press, 1973), p. 47.

34. F. Toennies, *Community and Society*, quoted in ibid., p. 46.

35. Ibid., p. 47.

36. See E. Morin, "La nature de la société," in *Communications*, vol. 22, (1974), pp. 3–32.

37. This is T.L. Thorson's leitmotive in *Biopolitics* (New York: Holt, Rinehart & Winston, 1971).

38. See S. Chodak, *Societal Development*, p. 48.

39. Ibid., pp. 47–48.

40. E. Morin, "La nature de la société," pp. 28–30; and *La méthode*, vol. 2, pp. 244–254.

41. A. Touraine, *The Self-Production of Society* (Chicago: The University of Chicago Press, 1977; translated by D. Coltman), Chapter 4 and *in passim*.

6

The Role and Functions of the Political System in the Polity

The political system performs several functions. One of these functions is to produce public goods (e.g., defense, lighthouses, etc.) since the market cannot organize their production and/or distribution. Considering the importance of the state-owned corporations in many industrialized or developing countries (e.g., Algeria, Canada, France, Italy, Mexico, the U.K., etc.) one finds it difficult to understand why Easton and Deutsch neglect the productive functions of political systems. This is perhaps due to the lack of direct involvement in manufacturing on the part of government in the United States; however, the paramount role played by the U.S. government in the "military industrial complex" considerably limits the significance of this formal non-interference. Easton's minimal interest in the productive function of public institutions is all the more unfortunate because these activities seem to conform with his model better than any other output of the political system. Furthermore, the political system as a producer of goods and services is, properly speaking, self-regulating but this is not necessarily the case for the political system considered as a "producer" of regulations and policies. (In the latter case, the "self" of the self-regulating system consists of the global society and not simply of the political system.) To explain these criticisms, I will first discuss the question of the boundaries of the political system and of the nature of its environment, before addressing the question of the scope of its activities, and examining the extent to which it is self-regulating or even self-organizing.

The Political System as a Societal Regulator

A unit of production transforms a wide range of inputs into a few outputs. W. Leontieff's input-output model of interindustrial relationships

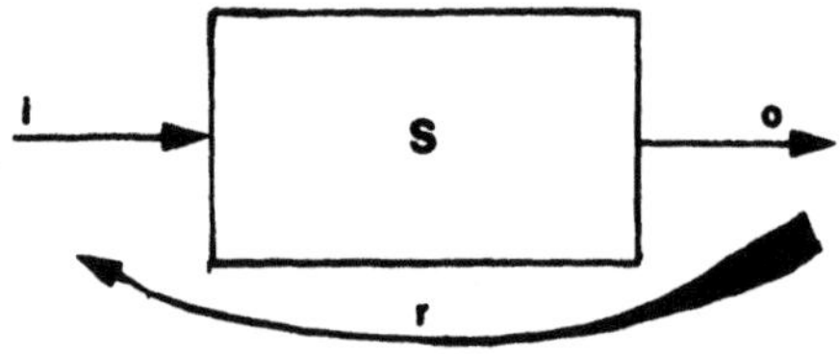

Figure 6.1

offer a clear representation of these flows. In this model there is no ambiguity about the boundaries of the system.[1] The environment of a given unit of production is the source of its inputs (e.g., raw materials); it also includes all the systems that need the outputs produced by this unit of production. The system itself is the space where the thermodynamic transformation of the inputs take place and where technological or organizational "know-how" is applied. It must be added that when the outputs react on the inputs, as is always the case to some extent in a finite world (e.g., automobiles use up the energy resources that are needed to produce them), one cannot define these various interdependencies as feedback processes unless the feedback channels have been clearly identified.

Despite Easton's protestation to the contrary,[2] his systemic model bears a striking resemblance to Leontieff's input-output model (see Fig. 6.1). Within the black box S it may be possible to identify a feedback loop transmitting information about the outputs to a regulator converting this information into instructions for the regulation of the input flow. If such is the case, the system is self-regulating (see Fig. 6.2).

It is important to stress that self-regulation is an *internal* process; the controlled funtion T and the controlling function R together describe the *same* system. This model adequately represents the structure of the pro-

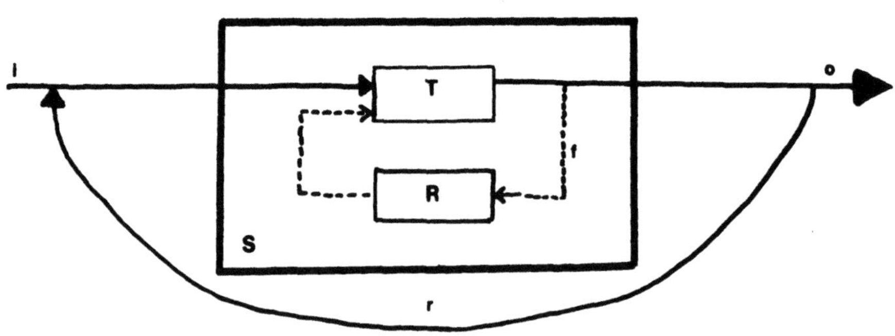

Figure 6.2

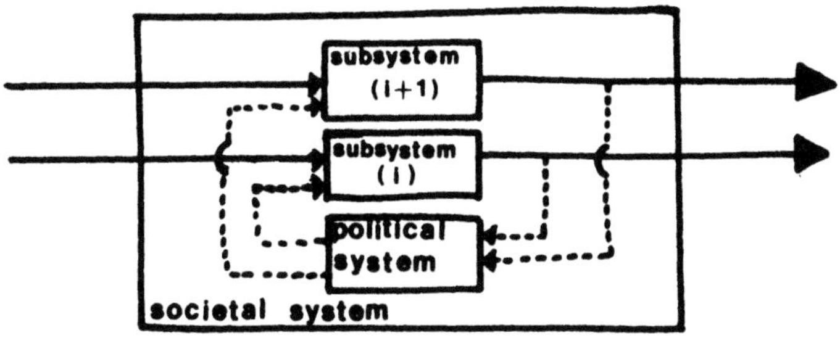

Figure 6.3

duction of public goods and services by a public sector which is capable of controlling its own productive activities but has no control over the usage of its outputs in an environment that looks very much like the market place.[3]

However, Easton's opinion notwithstanding, the "authoritative allocation of values" cannot be accounted for by means of an input-output model of the type outlined above. Easton's political system is really a societal regulator issuing commands, sanctioned by the use of power or legitimized by a recognized authority, to all the other societal subsystems. Yet he does not specify the boundaries, the functions, etc., of the controlled systems. This is a serious oversight, for these subsystems should not be confused with the "environment" surrounding both the political system and the systems it controls (see Fig. 6.3). Even though this model is already more empirically suggestive than Easton's model, insofar as it makes it imperative to identify and describe the concrete societal subsystems with which the political system interacts, it has little empirical value; for the analyst still has to disaggregate these systems, each one of which contains its own regulator.

As long as some important qualifications are added, Easton's definition of the political system as the most inclusive unit performing the "authoritative allocation of values" is consistent with the approach proposed here. First, as Easton himself concedes, the defining characteristic of the political system rests in its scope—i.e., that it is the most inclusive societal subsystem—as much as it rests in its regulatory function.[4] Second, except for its role in the production of goods and services, the political system does not allocate values simply to satisfy a demand for them. Its distributive and redistributive functions serve an ultimate purpose (e.g., to stimulate employment, or to promote regional development). Besides, it is often the case that the political system does not allocate values directly but rather imposes norms upon some controlled subsystem in order to constrain its allocative function (e.g., imposing credit restrictions on financial institutions). Third, because legitimation and coercion often occur simultaneously, Easton's almost exclusive preoccupation with relations of authority in the allocation of values must be counterbalanced by an equal concern with coercive practices. Yet these are distinct phenomena which cannot be reduced to one another; as Alain Touraine puts it:

> Institutions, because they are mechanisms of discussion, of negotiation, and of transactions, at the same time as being the political expression of a social domination, are always simultaneously instruments of *constraint* and of *legitimation*.[5]

Now, the definition of the political system presented above corresponds to an ideal-type. In reality, the political system hardly ever functions as a perfectly autonomous regulator. For instance, regulatory agencies are not immune to pressures from the very interests they are supposed to regulate (depending on a multiplicity of factors, including the national political culture, these agencies are more or less coopted by vested interests).[6] Besides, the political system does not control all aspects of all societal subsystems all the time—in any case, it would lack the requisite variety if it attempted to do so. The controlling activity of the political system is, in W.D. Coleman's words, "restricted to those facets of social life found useful for achieving particular consequences or functions."[7] Furthermore, there are situations in which a societal subsystem, e.g., the economy, can exert a controlling influence on the political regulators. For example, the labor market and the composition of the workforce broadly determine the quantitative, if not the qualitative, criteria of immigration policy.

The Scope of Political Regulation

Several authors adhere to the view that political life is, to a very large extent, a form of regulation. A critical review of some of the most significant theoretical frameworks built on such a premise will help to clarify my own position.

The subtitle of K. Deutsch's *The Nerves of Government* reads: "Models of Political Communication and Control." Thus Deutsch is more aware than Easton of the difference between the environment of the system *lato sensu* and the controlled systems about which information is fed back into the controlling, i.e., political, system; in fact, Deutsch often specifies the kind of policy field under consideration.[8] Yet he never explicitly raises the question of the identification of the variables that must be controlled in the environment. Is it important to improve upon Deutsch's concept of control? Are contemporary political problems and the analytical concerns of political scientists generally posed in terms of regulation? My contention is that modern politics is, to a very large extent, the politics of the regulatory process. Political science, as opposed to economics or even to sociology, has been rather slow in recognizing it. However, before discussing the contributions of political scientists to the analysis of the regulatory process, a brief review of Sir Geoffrey Vickers' work is in order.

Vickers, who writes in a refreshingly clear and simple style, has identified the dominant problems of advanced societies as stemming from a breakdown in the regulation of the ecology, the polity, and the psycho-social field.[9] Today this diagnosis sounds even more insightful in the face of the rise of what might be called the "false ecological consciousness" that blinds us to the long term effects of stopgap measures prompted by a worsening economic stituation (e.g., the substitution of coal for oil for the generation of electricity in the United States and Western Europe causes acid rain to fall on Canada and Scandinavia). Besides, most, if not all, post-modern political communities are torn by some form of "legitimation crisis" that extends beyond the public sphere into the private world of our disoriented consciences. Instead of vigorously reacting to developments, we simply try to "adapt" to them as if we still believed in the capacity of the market economy to regulate itself without external intervention. (There is indeed a considerable gap between our beliefs and our practices.)

Vickers defines regulation of the polity in broad terms as the ordering of "the relations between ourselves as doers and as 'done by'."[10] Thus a breakdown in the process inherent in political regulation would affect, in part, all the other modes of regulation. Vickers observes such a breakdown in the deepening of the contradictions between the two roles of "doers" and "done-by." He claims that the liberal worldview never developed a theory of "liberal conflict," i.e., one that would control conflicts through responsible political action, since "[t]he relation of enemy was supposed to have no place in world free from the tyranny of nature, men and gods."[11] The "post-liberal" era, when political regulation is again harnessing political power in order to meet the challenges of the environment and of the economy, will have as its foundation a renewed consensus evolved through a network of relational practices that can integrate shared experiences.[12]

Admittedly, Vickers' prognosis sounds rather vague, although not to the point of being meaningless. But I find very little to criticize in his understanding of politics as a regulatory process, except insofar as he even subsumes under this category the direct—as opposed to contracted and controlled—production of public goods and services. Obviously, one needs to distinguish primary from secondary processes in order to write meaningfully about regulation since it is, in an analytical sense and by definition, a secondary process.

Vickers' work has not received the attention it deserves from political scientists. However, the economist Gilles Paquet has recently outlined a suggestive "political economy framework" applying a kind of Vickersian cybernetics to the study of the political regulation of socio-economic processes.[13] Paquet's approach is in many respects consistent with the thesis defended here. Following Karl Polanyi, he considers "human economies" as complex networks of interdependent institutions and, following Vickers, he underlines their "relational/informational nature" and their "evolutionary character." From this macroscopic standpoint, he gives a rather positive evaluation of the effects of the state's growing involvement in the regulation of economic life. In particular, Paquet takes issue with most economists who are more sensitive to the short term inefficiencies induced by regulatory practices than to the long term benefits that accrue from a kind of learning process facilitating the adaptation of economic institutions to a turbulent environment. Paquet's characterization of the regulatory activities of the political system as a "moment" in a societal evolutionary process, rather than as an absolute evil (viz. the neo-classical economists) or as a panacea (viz. J.K. Galbraith), is similar to the thrust of this book.

One finds a similar cybernetic flavor in the works of Gérard Bergeron and of Vincent Lemieux.[14] The former makes an interesting distinction between the terms "milieu" (once very common amongst English-speaking sociologists) and "environment." For Bergeron the locus of politics (i.e., the "governance" of the public sphere) is the polity whose boundaries include the societal system. Such a perspective avoids the "unreal confrontation [with] a political system placed beside a social system, the latter being considered as the provider or environment of the former."[15] On the contrary,

> . . . the polity is structured by the very functioning of its governance. It bears in itself its society—being by definition a society—or rather its own social *milieu*. Its governance works within and not *upon* this milieu as if it were external to it.[16]

What has been stated above regarding the environment of a controlled system is fully consistent with Bergeron's well-taken point.

For Lemieux also politics is fundamentally a regulatory activity, and he understands regulation as a process of communicative interactions among a selector (or norms), a transducer (i.e., the subsystem that checks the variance between the actual and the desired conditions and determines the appropriate corrective measures) and an effector.[17] The latter is the controlled subject (or controlled subsystem); the fact that the controlled entity is explicitly included in the network by Lemieux vindicates, again, the distinction made earlier between simple input-output model and a cybernetic model of controlled production through a network of communicative interactions. Lemieux maintains that, in such a network, actors can play multiple roles. Thus one can be a subject (i.e., an effector) at one time and an "agent" (i.e., a part of a transducer) the next time around. Moreover, he distinguishes between "cybernetic roles" that are defined by the actor's function in the transmission of information, and "positions of influence" that are defined by an actor's ability to control how information is processed in the network and the uses to which it is put.

According to Lemieux, cybernetics ought to be one of the pillars of political science but not the only one. It is ideally suited to the macroscopic level of analysis but it cannot account for phenomena that occur at the level of interpersonal or intergroup relations, nor can it account for the structural logic that is realized in a given communication network. The former, according to Lemieux, who acknowledges his debt to M. Crozier,[18] can best be conceptualized as competing strategies and, therefore, lend themselves to a game-theoretic approach; the latter calls for an adaptation of structuralism to political science. To conclude, Lemieux's original synthesis of cybernetics, game theory and structuralism constitutes one of the most sophisticated theoretical interpretations of political regulation. Unfortunately, it remains strictly synchronic throughout, and, while it helps us to understand a crucial "moment" in the reproduction of political life, it does not address that question directly.

So far I have discussed societal regulation as if it were the more or less dominant concern of state authorities. This is entirely reasonable from a cybernetic point of view. It is also an acceptable hypothesis from an historical perspective. For example, even the non-interventionist state posited by orthodox liberalism indirectly regulates economic transactions insofar as it legitimizes and protects the self-regulating market economy, which is itself a societal arrangement no more "natural" than the mercantialist system that it replaced. However, it may at times be necessary to distinguish those public policies that are unequivocably regulatory from those that are not. In this respect, Theodore Lowi's typology is most useful.

Differentiating policy types in terms of the applicability and the likelihood of coercion, Lowi identifies four categories: distributive policies (e.g., tariffs and subsidies), constituent policies (e.g., setting up a new agency, propaganda), regulative policies (e.g., elimination of substandard goods, unfair competition or fraudulent advertising), and redistributive policies (e.g., progressive income tax, social security), as Table 6.1 illustrates.[19]

Much of the literature on policy-making takes it for granted that the socio-political environment conditions the nature of the policy response. By contrast, Lowi suggests that the opposite relation is also worth exploring. That is, the nature of the policy under consideration has itself a determining influence on the political strategies of the decision-makers. For example, the "Republican era" (1869–1901) in the United States was characterized by congressional dominance and presidential passivity precisely because, according to Lowi, federal politics at that time was concerned primarily

		Applicability of coercion (works through)		
		Individual conduct	Environment of conduct	
Likelihood of Coercion	Remote	DISTRIBUTIVE POLICY	CONSTITUENT POLICY	Party logrolling vs. Group bargaining
		REGULATIVE POLICY	REDISTRIBUTIVE POLICY	
		Decentralization vs.	Centralization	

Table 6.1: Lowi's Typology

with distributive issues (e.g., land disposal programs), and distributive policies are worked out through many trade-offs among a variety of individual actors, such as is usually the case in congressional committees. Redistributive policies, on the other hand, divide the polity along class lines and their successful implementation requires the kind of strong leadership that Franklin D. Roosevelt is remembered for. But Roosevelt also knew how to accommodate and cajole pressure groups concerned about regulatory issues, as these involve yet another style of conflict resolution. Consequently, if one believes that important policy issues ought to be discussed in the political arena in a manner that allows for a significant degree of participation on the part of organized groups or the interested public, Lowi argues that distributive policies ought to be avoided and regulative or redistributive policies are to be preferred.

What emerges from this review of apparently unrelated models of policy-making is the concept of the autonomy of the state. Cybernetics emphasizes the autonomy of the regulator (i.e., governmental institutions) vis-a-vis the controlled societal conditions. Both Lemieux and Lowi, on the other hand, remind us of the fact that policy-making is not a mechanical process. Individual actors pursue their own strategies within the institutional constraints that they face; these constraints can themselves become the object of political decisions, as when a choice is made between alternative policy instruments, that is alternative modes of political interaction. This perspective differs significantly from much of the contemporary political science literature which is "society-centered" rather than "state-centered," to use E. Nordlinger's terms.[20]

Self-Regulation and Self-Organization of the Political Regulator

Thanks to its political system a given society is self-regulated; but to what extent is the political system itself capable of regulating its own activities: *quis custodiet custodes*? The position adopted here is that only the political community as a whole—as opposed to its political system—is fully self-regulating and, moreover, self-organizing (i.e., capable of creating the conditions of its own regulation). However, the contrast thereby established between a functionally dependent regulator evolved over a long period of time through a combination of societal processes extending beyond its scope and incapable of monitoring its own activities, and, on the other hand, a fully autonomous polity needs to be relativized. This over-simplified characterization represents only a first approximation. Again, the very notion of complexity entails the existence of partially autonomous subsystem within anything called a complex system; thus, I shall hereafter attempt to evaluate the margin of autonomy that a political system commands.

David Easton coined the terms "withinput" and "intrasystem output" in order to describe the self-regulatory capabilities of the political system.[21] By acknowledging that withinputs may be destablizing, Easton seems even willing to stretch the meaning of these terms in such a manner as to come rather close to proposing an autopoietic interpretation of political life:

> What political systems as a type of social system possess uniquely . . . is the capacity to transform themselves, their goals, their practices, and the very structures of their internal organization. To keep the vital processes, the essential variables, of a political system alive, as it were, a system must remodel its structures and processes to the point where they are unrecognizable.[22]

Unfortunately, Easton does not elaborate upon the conditions for the realization of these internal transformations.

Assuming that the political system of a given plural, differentiated society is able to transform itself to some extent—although never to the full extent of being able to reproduce itself, since it does not control all the societal resources required for its complete reproduction—I shall here again use experience, consciousness and self-consciousness as the three dimensions of an analysis of the internal dynamics of that system.

First, political life amounts to the production of information-as-command (or "imperative information" in Lemieux's terms). There is no point in restating here Lemieux's analysis of the competing strategies deviced by actors in "positions of influence" within a cybernetic network. Let us simply reflect upon the significance of this competition with reference to the order from noise principle. Clearly, those who control internal communications within the transducer, even if they do not initiate messages, are in a position to bring about political crises by closing existing channels or opening new ones, and then to resolve these disturbances to their advantage. They contribute thereby to the self-organization (or simply self-regulation depending on the import of the changes involved) of the political system in which they operate. For example, K. Deutsch considers "middle-level groups" (e.g., colonels in an army) as the agents most likely to set in motion political changes (e.g., military coups).[23]

Second, one can also observe within any political system tensions developing between, on the one hand, our aspirations and the defenses we build around them and, on the other hand, societal order. These problems are very numerous and can be approached from a variety of perspectives.[24] Two such problems are most directly relevant to the study of policy-making. The first one concerns the decision-makers' ability to recreate their own world and thus to adapt to complex situations; the second one is almost diametrically opposite and concerns the human tendency to shield oneself from external threats by protecting one's own world—a tendency that seems to characterize bureaucratic behaviors in particular. Both these socio-psychological phenomena may turn out to be the source of organizational crises.

J.D. Steinbruner claims that the assumptions of "analytic" decision-making theory according to which individuals carefully weigh the costs and benefits associated with the probalistically distributed outcomes of alternative choices does not hold under conditions of high uncertainty. He suggests instead that decision-makers tend to organize their uncertain environment by the practice of "categorical inference."[25] Although such a faculty usually permits decision-makers to solve the problems confronting them, it may also trigger destabilizing fluctuations and thus aggravate the complexity of their environment. One of the assumptions of cognitive theory, explains Steinbruner, is that subjects try to preserve the consistency of their thought-patterns; thus new information will be interpreted so that it does not contradict established beliefs—he calls this "inconsistency management." One aspect of this process, namely wishful thinking, clearly brings into play the nebulous part of the human mind where imaginary creations disturb the fragile order built by human reasons. In Steinbruner's words:

> When a set of beliefs is under pressure from inconsistent information being processed in a short time frame, it is possible to maintain consistency without changing the beliefs by casting them in a long time frame and adapting the inference of transformation; namely that the immediate situation will succumb to a favorable trend over time.[26]

To illustrate this point the author gives the example of the American military operations in Viet Nam in the early 1960s; Prime Minister Trudeau's attitude towards separatism in Quebec during the 1970s is another case in point. Both these examples suggest that the tendency to believe in one's own illusions accelerates the pace of events, thereby rendering "catastrophic" (in R. Thom's sense of the term) transformations of the political system more probable.

The mental framework of bureaucrats constitutes another instance of destabilizing "false consciousness." Michel Crozier's well-known analysis of the French bureaucratic behavior provides us with an instructive example.[27] The characteristic avoidance of face-to-face communications across hierarchical echelons results in the implementation of rigid, impersonal rules that can only be altered through periodic cathartic crises. This may be interpreted as a culturally specific trait, but R.P. Hummel claims that the pursuit of self-denial and anonymity is a psychological attribute of bureaucratic behavior *per se*.[28] Organizational sclerosis strikes the political system because it is largely composed of depersonalized bureaucrats incapable of entering into the kind of interactive relations that could ensure prompt adaptive responses to changing environmental circumstances. Hummel alleges that interpersonal relations in the bureaucratic world are based on psychological manipulations (viz. the fear of losing one's identity) rather than on legitimation. Therefore, the bureaucratization of politics that so many observers have denounced logically leads to a weakening of relations of authority based on the adherence to some "transcendental" meaning. But Hummel does not postulate that such a trend would result in a highly unstable political equilibrium were it to continue. On the contrary, I concur with Jean Blondel's thesis that a purely administrative state would be likely to create, over time, the very conditions of its repolitization, precisely because its legitimacy would always remain in doubt.[29]

Eventually, the third dimension of the internal self-organizing dynamics observable in political systems consists in the rational search for the "best" form of government. One may object that attempts to model political institutions and practices on some ideal notion of the good must be envisaged in a broad historical and societal perspective. In the long run, this is indeed the only valid position. Again, the central theme of this chapter is that the process of political self-organization cannot be fully understood as a process that would be limited strictly to the boundaries of the political system. For this reason, I will return to the problem of establishing an ideal political order in the next two chapters. However, this does not preclude the possibility of observing nascent self-organizing tendencies within the political system.

Practically, this means comparing the respective contributions to institutional design of political experience and technocratic expertise. During the last two or three decades we have witnessed the emergence of a decisional technology (e.g., the Planning-Programming-Budgeting System) which has radically altered the context, if not the substance, of public policy-making. These techniques challenge traditional forms of political accountability and represent a good example of the kind of "noise" that can develop in a self-organizing system.

However, the political system remains too dependent on the culture of the society in which it operates to be able to reproduce itself entirely. While the system cannot be compared to a servomechanism conceived by an external designer, since it is partially capable of organizing itself, it still cannot produce all the conditions of its own production. Societal reproduction is a global process but one in which politics plays a central role.

Notes

1. The point here is not that it is always easy to delineate these boundaries, but rather that the difficulties one may reasonably expect to encounter will be of a technical rather than epistemological nature.

2. *A Framework for Political Analysis* (Englewood Cliffs, N.J.: Prentice-Hall, 1965), pp. xi–xii.

3. But, of course, when the goods are "public goods" in the strictest sense of the term according to the theory of welfare economics, the market is by-passed.

4. I do not share Vincent Lemieux's "extensive" definition of politics as something characteristic of all human organizations insofar as they are capable of maintaining their organization for reasons that were presented above (i.e., politics is a societal, public, global process). Obviously, there are important homological correspondences between private and public organizations, or between kinship and society but, from a dichronic perspective, one cannot ignore the equally significant differences that separate these levels of social reality. On the other hand, I share many methodological assumptions with Lemieux, as it will become clearer below. See Vincent Lemieux, *Les cheminements de l'influence* (Québec: Les Presses de l'Université Laval, 1979), pp. 4ff.

5. A. Touraine, *The Self-production of Society* (Chicago: The University of Chicago Press, 1977), p. 178.

6. In countries such as Canada where the industrial and commercial operations of the state are important, the "vested interests" can themselves be publicly owned corporations (e.g., the relations between the Canadian Transport Commission and Air Canada or Canadian National Railway).

7. W.D. Coleman, "Political Demands and Political Structure: Some Suggestions Relevant to a System Analytic View of Political Change," paper presented to Annual meeting of the Canadian Political Science Association, May 1978, p. 4.

8. Deutsch's elaborate diagram of the political system (*The Nerves*, p. 258) is meant to illustrate more specifically the foreign policy-making process.

9. G. Vickers, *Freedom in a Rocking Boat* (Penguin Books, 1972).

10. Ibid., pp, 56–67 and 155.

11. Ibid., p. 191.

12. There are interesting parallels between Vickers' ideas and Etzioni's attention to societal consensus-building as a prerequisite for the realization of societal changes corresponding to the needs of a large majority of societal actors (*The Active Society*, Part Four). But Vickers appears more concerned with the protection of the minorities and of the individuals; indeed his "post-liberal" era promises to create the conditions under which the liberal dream would come true—this, Vickers understands, is an

essentially political problem, whereas early liberals thought that the main political problem was how to avoid politics altogether.

13. G. Paquet, "The Regulatory Process and Economic Performance," in G.G. Doern (ed.), *The Regulatory Process in Canada* (Toronto: Macmillan, 1978), pp. 34–67.

14. Gérard Bergeron, *La gouverne politique* (Québec: Les Presses de l'Université Laval, 1977), and Vincent Lemieux, *Les cheminements.*

15. *La gouverne*, p. 16 (my translation).

16. Ibid., p. 17 (author's emphasis; my translation).

17. Although such a terminology is quite traditional, Lemieux's work contains many interesting and original ideas on cybernetics proper and on its relevance to political inquiry. In particular, his emphasis upon the fact that regulation supposes the transmission of a special kind of information, namely, "imperative information" as opposed to ordinary "indicative information" (*Les cheminements*, p. 33) echoes the developments on "information-as-command" found in Chapter 2 of this book. Upon this concept he elaborates a definition of "informational power" the effects of which amounts to a form of "instigation" and contrasts it with "energetic power," i.e., the exercise of coercion.

18. M. Crozier and E. Friedberg, *L'Acteur et le système* (Paris: Seuil, 1977), *in passim.*

19. T. Lowi, "Four Systems of Policy, Politics and Choice," *Public Administration Review*, July/August 1972, pp. 298–310.

20. Eric Nordlinger, *On the Autonomy of the Democratic State* (Cambridge, Mass.: Harvard University Press, 1981), p. 1.

21. *A Framework*, pp. 114–115, and *A System Analysis of Political Life* (New York: John Wiley, 1965), pp. 55–56, 347–348.

22. *A Framework*, p. 99.

23. K. Deutsch, *The Nerves*, pp. 154–155.

24. See for example G. Wallas, *Human Nature in Politics* (1908); H. Lasswell, *Psychopathology in Politics* (1928); T.W. Adorno *et al.*, *The Authoritarian Personality* (New York: Harper & Row, 1959).

25. John D. Steinbruner, *The Cybernetic Theory of Decision: New Dimensions of Political Analysis* (Princeton, N.J.: Princeton University Press, 1974), p. 110.

26. Ibid., p. 117.

27. M. Crozier, *The Bureaucratic Phenomenon* (Chicago: University of Chicago Press, 1964).

28. Ralph P. Hummel, *The Bureaucratic Experience* (New York: St. Martin's Press, 1977); note that for Crozier bureaucratic practices are the *result* of the various strategies deviced by free individuals within the system who attempt to preserve their identities, for Hummel, on the contrary, these practices have a destructive impact upon the self-image of bureaucrats.

29. Jean Blondel, *An Introduction to Comparative Government* (New York: Praeger, 1969), pp. 405–412.

7

The Production and Reproduction of the Nation-State

An approach equating political life with the production and reproduction of a nation-state is *prima facie* fraught with conceptual difficulties and charged with dubious ideological connotations. To posit the existence of something like a national psyche, as I do, must sound exceedingly idealistic to all but the most romantic political theorists. Moreover, to speak of a nation-state is to use two rather indefinite terms whose conjunction may even be problematical in some cases.

I argue hereafter that the concept of nation has at least two meanings which are rooted in different historical traditions; one roughly corresponds to the conscious, and the other to the superconscious dimensions of societal order. Each one of them may lead to some historical aberration, if it is allowed to develop independently of the other. The intended objective of this discussion is not to propose a comprehensive analysis of nationalism—something which has already been done.[1] It is rather to illustrate that the creative function of disorder, when it involves relations of power, can lead to catastrophic disruptions of the evolutionary process, i.e., to destructive, plain and simple disorder. And yet very few polities can exist as something other than as nations, and thus completely escape the pull of nationalism.

Consequently, the study of political culture is much more than a specialized subfield of the discipline: it is a prerequisite to political inquiry. One of the objectives of this chapter is to propose some hypotheses and to discuss some crucial methodological problems concerning the analysis of national political cultures.

I also argue that, even though not all nations are fully sovereign nor are all states coextensive with a nation, and even though statism is one of the most odious ideologies of the modern era, the state continues to be an indispensable constituent of a self-organizing polity. Whereas national myths contribute to the formation of a collective identity, i.e., to the formation of the "self" in societal *self*-organization, the state forms the organizing matrix of that self. The state is envisaged here, not simply as a mechanistically functional apparatus, but also as an object of (unequal) exchanges of symbols. The nation-state still represents the archetypical polity. This means that political self-organization revolves around the material and semiological production and reproduction of a nation-state, although it cannot be reduced to it. Consequently, the subject matter of a science of politics looks very much like that of traditional political philosophy: it concerns the production, the implementation, and the creative reproduction of ideas of the good within a politically organized and sovereign community.

The Nation as the "Self" of the Polity

Ever since the French Revolution, fascinated or horrified political thinkers have reflected upon nationalism. Today, many are those who claim that the time has come to look beyond national barriers.[2] I agree, but no matter how distasteful nationalistic fervour may be to many political scientists, it is imperative not to let one's preferences color reality to the extent that it becomes unrecognizable. Unless we understand fully the significance of the national phenomenon (and of its persistence), and unless we are prepared to make room for it in our models of political life, there is little hope that we shall succeed in transcending it. This type of inquiry could perhaps yield some elements of a solution to the Habermasian problem of understanding how contingent historical events come to bear upon the developmental logic of world-views and of morality.

The ardent but hazy ideal of the Jacobins was pregnant with two images of the nation. One, which was to become more and more important during the revolutionary wars, emphasized the defence of the French territory and of the cultural heritage; it was inherently intolerant of local particularisms and of languages other than French (viz. the germanic dialect spoken in Alsace, or the Celtic Breton language). Ironically, this ethno-centric tendency was to give rise in French-occupied Germany to Fichte's romantic belief in the existence of a "German Spirit" likened to "an eagle whose mighty body thrusts itself on high."[3] The irrationality of this kind of exalted discourse has been carried through the German nationalist tradition which, until the recent past, has tended to emphasize criteria of nationhood such as the uniqueness of a given land, history, language and even, tragically, "race." The allegiance of those who form the nation is posed *a priori* as being natural and therefore unproblematical. It has been called the religion of *Kultur*. However, it would be wrong to imply that such an ideology has always been, and continues to be, a peculiarly German tradition. Maurice Barrès' nationalist doctrine is built upon very similar principles, and French

Canadian nationalism from the repression of the rebellion led by Papineau' Patriotes in 1837 until the Quiet Revolution of the 1960s was similarly devoid of voluntaristic justifications.

However, there is another aspect of the Jacobins' nationalist discourse which also needs to be taken into account. This was a paradoxically universalist and voluntarist conception of the French nation as the vanguard of democracy and as the missionary of a new faith in the rights of man to which her revolutionary armies tried to convert the rest of Europe. Upholding this tradition, Ernest Renan denounced the serious mistake of confusing "race" with nation in his pamphlet *Qu'est-ce qu'une nation?* written in 1888. Like Fichte, Renan defined a nation as a "spiritual principle" but one that is firmly grounded in a will to live in common—for Renan, the nation is the outcome of a "daily plebiscite." Let us briefly recall the historical context in which this was written. After 1871 Germany established its claim over Alsace-Lorraine on the basis of the Germanic character of these provinces; France maintained that the expressed desire of the native population to remain a part of the French nation invalidated that claim. In the 20th century, Renan's voluntaristic conception of the nation, albeit stripped of any reference to a "spiritual principle," gained a considerable influence. In 1919, it was embodied in the Wilson doctrine which led to the dismemberment of the Austro-Hungarian empire; it later inspired all the liberation movements of the decolonization era.

Neither interpretation is satisfying in that Fichte ignored the fact that man is both the product and the maker of his societal milieu. Renan, on the other hand, seemed to imply that any individual or group could, at any time, on any issue, withhold consent—a prospect which, as D. Cameron aptly remarks,[4] is unlikely even in the best of circumstances. And yet these doctrines have been very influential, sometimes in perverse ways. Indeed the odious effects of some forms of nationalism ought to remind us that the paradoxically creative implications of societal crises are not always concretely realized. The excesses attributable to ethno-centric nationalism, and to racism in general, are all too well known. Today, it seems more urgent to consider the threat posed by the more voluntaristic and rationalist variants of nationalism, even if it may be a somewhat less serious one.

After the Second World War, and for obvious reasons, internationalism became a very attractive alternative to nationalism; it inspired such momentous entreprises as the United Nations and the various attempts to unify Europe. The rapid spread of technology throughout the world, the growth of multinational corporations and the polarization of the international scene, to mention only a few factors, seemed to render traditional historical communities obsolete. But it is almost as if the "current of history" took a sharp turn in the opposite direction: during the last two decades scores of new nations (i.e., nation-states) have won their independence from the colonial powers and now, in old as well as in new nations, many ethnic groups, or rather militant groups therein, are demanding the right to self-determination.

Actually, both developments partake of the same logic. In both cases the supposed rationality of the goal being pursued justifies a willingness to engage in mystification. By will, and by will alone, rational men thought that they could, for example, federate Europe or make the United Nations into an instrument of collective security. However, after the internationalist movement came to experience what Jean Jaurès called "the incoherent slowness of universal effort,"[5] the stage was set for the emergence of a new kind of nationalism: the ideology of ethnic liberation.

One of the most intriguing themes of the neo-nationalist ideology is that the formation of many new small nations will destroy the oppressive bureaucracies developed by existing nation-states and by their supranational off-springs.[6] But it would be wrong to conclude that the separatist and neo-nationalist movements, which have become so visible in many countries in Europe and elsewhere, are primarily inspired by an irrational and radical rejection of modernity, as it had been often the case of similar movements in the first half of this century. Admittedly, we are dealing here with a complex and multiform ideological tendency that cannot be reduced to a single dimension.[7] However, it seems that a pattern of politicization of intelligentsias preoccupied with the structural decline of regional economies and with the socio-economic deprivation endured by the ethnic groups with which these intelligentsias identify is common to most separatist movements. Characteristically, modernization is then presented as a compelling goal that can be achieved only through self-determination. The most noteworthy aspect of these neo-nationalist movements is perhaps the rationalistic nature of a discourse that carries the ideas of self-determination and self-government to an often absurd limit. Considering, for example, the immense distance that separates Corsican activists from the vast majority of the residents of Corsica (and the same could be said of the situation in Wales, Scotland or even, with some qualifications, in the Punjab), one is led to the conclusion that many neo-nationalist movements aim at a specious kind of self-determination which the national "self" is not collectively experienced but simply willed by a political/cultural elite. This might be a premature conclusion in some other cases like that of the French-speaking *Québécois*—but even in this case, the population of Quebec unequivocally rejected the "sovereignty-association" option proposed by Premier Lévesque in May of 1980.

The point here is not that all demands for regional and/or cultural autonomy are illegitimate; it is rather that many contemporary variants of nationalism (or of internationalism, for that matter) rest upon mystifying ideological concepts that do not reflect the concrete historical realities of the national communities in question. Precisely, I argue later on that the complexity inherent in self-organizing structures like national or ethnic communities are not easily translated into the logical categories of traditional ideological discourse.

As for romantic (often simply xenophobic) nationalism, there is little need to dwell upon the tragic consequences that it entails. It feeds upon and expresses deep-seated fantasms through which the national psyche is connected with the private fears of the individual members of the national community. Henri Atlan, for example, notes that there is something incestuous about the cult of the land and of the dead, since the nation seeks thereby to possess that which gave it birth.[8] But such a nationalism proceeds from premises that are no less potentially universal, and thus no less rational than the idea of self-determination. It is entirely rational for the members of a long established political community to value highly the preservation of a way of life which they can experience as being the unique product of their ordinary life and that of their ancestors in a well-defined territory. By definition, an individual or a collective identity is unique. To preserve and nurture one's identity is a strategy that the right to pursue one's own purposes, creed or destiny fully justifies. However, the substantial rationality of this historical destiny, and of the individual sacrifices that its realization entails, ought never to be assumed *a priori*. Such was, though, the serious mistake made by Fichte for whom the German nation's superiority was preordained; Hegel went even further, stating that the Prussian state was the outcome of the dialectical evolution of pure reason, God on earth!

At this point, we can come to some preliminary conclusions. Very few politically organized communities can exist as anything else than as nations and maintain their autonomy. Therefore, the nation appears to be an "evolutionary universal." Insofar as a national myth constitutes the "self" of a self-organizing polity, it delineates the symbolic space within which political interactions take place. National solidarity is in many respects an affectional pattern even if, as we have seen, its inception owes much to the rationalism of the Enlightenment. Indeed, those aspects of nationalism that I previously interpreted as having been derived from rational premises perhaps ought to be conceptualized as instances of a (non-rational) valuation of reason in the formation of national myths rather than as the use of reason itself. On the other hand, it is not impossible, and it is certainly desirable, to maintain a distinction between the idea of nation and nationalism even if, in the history of ideas, the former was developed at about the same time as the latter. In order to justify such a distinction, it is crucial to understand that it is based on a recognition of the differences separating practical from reflexive political consciousness. On the plane of practical consciousness, the nation remains a relatively amorphous and typically paradoxical idea derived directly from experience; it is likely to include references to both universal and particular (i.e., historically situated) elements of the human condition, as was the case of the Jacobins' discourse on the French nation. The paradigm of the new cybernetics permits us to deal with such paradoxes, since it posits that morphogenetic processes are to be analyzed from a dialectic—or, to use Morin's term, a "dia-logical"—perspective. Political actors faced with practical problems such as the definition of their national identity are also usually able to sort them out to their own satisfaction. They may, for example, choose not to make explicit the criteria of nationhood (at least in ordinary language), thereby allowing for an endless series of operations of inclusion or exclusion determined by changing circumstances. However, in periods of crises, it may become difficult for the governing authorities—or for revolutionary forces—to mobilize support for—or opposition against—the regime or the political community as a whole. An undetermined, implicit and unarticulated representation of the national identity is no longer acceptable to those who seek to reshape its image. Nationalist ideologies are then used to operate a reductionist transformation on the national phenomenon (e.g., the nation as a purely rationalist or romantic construct). The nation then becomes the distorted image of the ideological values and priorities of those whose strategic position in the political arena gives them the power to operate such a transformation. Admittedly, the national psyche always reflect, to some extent, the political and cultural elites' values. But the point here is that ideological discourse applies a deforming logical structure to reality (e.g., the pseudo-Wagnerian imagery inherent in Nazi rituals). Therefore, nationalist doctrines fail to express the complex meanings of the symbolic representations of the nation that are vehicled through ordinary language and are re-interpreted by everyone in the national community through redundant cultural codes and personal experiences.

To stress that a national being is formed when a network of interpersonal communications is being used to transmit messages about shared political and cultural experiences does not mean, however, that the national phenomenon cannot be the focus of a critical reflection, nor that it must escape normative judgments. So far I have been concerned with the less than truly analytical uses of human reason in the formulation of nationalist ideologies and I have underlined the tragic consequences of nationalism. The question now is: Can we think of better ones?

A salutary nationalism should, on the one hand, serve a community in achieving a sense of self without which there can be no meaningful collective endeavor and, on the other hand, leave open the possibility of reaching another stage in societal evolution (presumably a more universalistic one). In other words, it ought to be both practically relevant and yet sufficiently ambiguous to remain a topic of diverging interpretations among societal actors. It would create tensions and, at the same time, set limits to their development.

Could not Renan's original (in both senses of the term) definition which, let us recall, combined in a somewhat perplexing way rational will with a spiritual principle, precisely represent the kind of creative polarity required to transcend the two main currents of nationalistic thought? It would seem that such a reflective ambiguity would be particularly relevant to the citizens of multi-cultural countries (e.g., Belgium, Canada, India) insofar as it might help to produce a meta-nation out of the existing communities, not by relying on will alone nor by means of an impossible rewriting of history, but through the sharing of a common determination to preserve the "spiritual" uniqueness of the constituent national identities.

National myths are not the only means of achieving solidarity within a polity. Regional loyalties, urban or rural communities, parties and interest groups perform a similar integrative function, albeit a different level. They too are produced and reproduced through political interactions. While these alternative constituents of a collective self are not necessarily incompatible with national solidarity, there is another type of solidarity that some would regard as antinomic indeed, namely class consciousness. Social classes are an undeniable reality. In fact, they have a quasi-biological nature: one is born into a class just as much as one is born male or female and most, although not all, individuals remain in the same class throughout their lives. Thus the vocabulary or reproduction is particularly well adapted to the analysis of these phenomena, as it is indeed illustrated by Marxist theory.[9] But when a conflict arises between class loyalties and national solidarity, the latter tends to prevail, as the European socialist parties painfully discovered at the outbreak of the First World War. Admittedly, historical examples of "class collaboration" in the face of external aggression may not be sufficient to refute the primacy of class consciousness and class interests in the reproduction of the polity; the unity of the latter could simply be an artificial projection upon society as a whole of the commonality of interest binding the ruling class. The debate remains open. However, this debate does not bear upon the central position put forward in this chapter. The point at stake here is that a properly understood cybernetic approach to politics, that is to say, one that does not reduce cybernetic theory to its technological applications, should not be limited to the investigation of the functional aspects of political communications. It ought to raise explicitly the question of the identification of the totality that emerges from these interactions and renders their rearrangement possible by, paradoxically, ordering their destructuration. I have chosen to illustrate this point by means of a brief discussion of the importance of national myths. It could, although I think less probably, be defended in terms of an analysis of class formations.

The second observation that can be inferred from an identification of the self of a given polity with a national psyche is addressed to political scientists as participant observers of their own polity. If national myths enter so prominently in the definition of their subject matter, the study of national political cultures does not simply constitute a specialized subfield of political inquiry but is rather one of the necessary conditions for its practice. Actually, this is beginning to be acknowledged now. However, if national myths can only be understood in relation to the ideational complexes formed by national cultures and in relation to history, then it follows that the study of political cultures requires the adoption of a politically significant and *comprehensive* perspective.[10] In other words, the study of political cultures should be characterized more by its political relevance than by its specific focus.

Some Paradoxical Research Hypotheses on Political Culture

We live in a complex world that we would like desperately to believe is simpler than it really is, hence the success of "demystifying" ideas. Political scientists too often act as exterminators bent on destroying the mythical "monsters" that stand in the way of our mental welfare. What follows is a modest attempt to rescue some of the paradoxes relating to the study of political culture which competing schools of thought have dismissed. My aim here is not so much to make a significant contribution to the study of political culture as to illustrate what "monster conservation" entails in view of the preceding analysis of the paradoxical character of political life.

Two problems will be discussed:[11] the relationship between the concepts of political culture and ideology, and the relationship between the "micro" and "macro" levels of analysis.

Culture or Ideology?

The first problem is that of the definition of political culture itself. Is political culture a subset of the generic concept of ideology, or is it the other way around? I want to argue that these two positions are both paradoxically true.

The concept of political culture was originally proposed by G. Almond and others in order to circumvent ideology by focusing on the unsystematically articulated "orientations" that members of a political system share across ideological lines. In this sense, a political culture encompasses several ideologies. As Bell and Tepperman explain, this perspective suffers from two weaknesses: it cannot account for the socio-economic foundation of cultural values, nor can it account for cultural changes over time.[12] The

Marxian notion of ideology appears to be more satisfactory in this respect; thus ideology would be the primordial, originating world-view and political culture a derivative notion. However, were we to accept this position, we would be unable to understand the peculiar symbiosis that occurs when, over a long period of time, members of a polity reproduce cherished ideological conflicts. The autonomy of the autopoietic polity finds its expression in a national psyche which, as we have seen, assumes a life of its own.

One possible way of coping with this paradox is to make a distinction between "values" and "norms." For G. Vickers,[13] values are "abstract words of great ambiguity and imprecision." Precisely, this ambiguity, coupled with the fact that values often "come in complementary and partly inconsistent pairs" (e.g., freedom and order, justice and mercy, etc.), make it necessary for each generation to discuss and redefine their content. Therefore, "values are general and explicit." By contrast, "norms are specific and tacit." They are the standards to which political actors refer in order to judge the acceptability of a concrete situation (e.g., standards of public health, education, institutional performance, morality, etc.). For the purpose of the present discussion, I regard values as the components of ideologies and norms as those of (political) cultures. Great care should be taken in empirical studies to distinguish these concepts which must be apprehended by different techniques (e.g., content analysis of speeches, platforms, etc., as far as the identification of values is concerned; questionnaires, "games," etc. as far as norms are concerned). However, this is an insufficient description of the complex ideational dimension of political life. We also have to take into account the interactions whereby values and norms are evolved and transformed. I agree with Vickers that these transformations result from the operation of recursive processes but I shall depart from his analysis hereafter.

The paradoxical identity of opposites at stake here involves a dialectical tension between the consensual and implicit character of both the "national psyche" and Almond's political culture and, on the other hand, the conflicting and explicit ideologies. As a first approximation, Figure 7.1 schematizes these relationships. On both the left and the right sides the representations of societal reality are overlapping to a very large extent (although never fused) whereas ideologies are distinct (but they may form clusters of partially overlapping ideologies, e.g., social democracy and Marxism-Leninism).

The rationale for the inclusion of the "space of archetypes" in my model is that I do not consider that Vicker's "values" are the most encompassing valuational referents; moreover, the symmetry is needed to account for the paradoxical character of ideologies which either subsume consensual representations or are subsumed by them. The investigation of the space of archetypes is best undertaken by means of historical methods of analysis (as the diagram suggests, interacting ideologies are transformed into archetypes

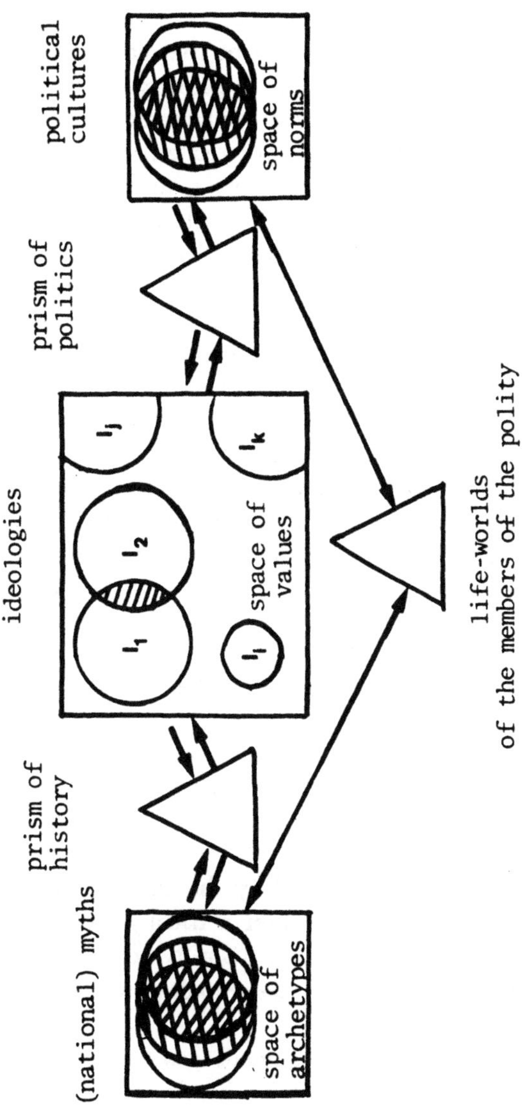
(national) myths
space of archetypes
prism of history
ideologies
I_1
I_2
I_j
I_k
I_l
space of values
prism of politics
political cultures
space of norms
life-worlds
of the members of the polity

Figure 7.1

and articulated into national myths through the "prism of history" by which I mean that diachronic and concrete process of societal autopoiesis). Among these historical methods, philosophical hermeneutics stands out as the most appropriate one in this instance, but other methods can be used also, such as psychoanalytic introspection—indeed I deliberately used the Jungian term of archetype to characterize the left hand side referents.

I must now turn to the right hand side of the diagram for it is there that the merits of a perspective that reflects the paradoxicalness of complexity appears most clearly. The power structure built in the "prism of politics" allows one ideology or, at most, a cluster of related ideologies to become dominant and then to be converted into norms, in the Vickersian sense of the term. This conversion process is particularly complex for it involves, not only what Parsons calls the "internalization of values," but also instances of psychic coercion whose effects take the form of a grudging, cynical or humorous acceptance of the norms. The latter mode of norm creation, which A. Giddens has cogently analyzed,[14] illustrates the significance of the phrase *dominant* ideology. Critical or marginal ideologies do not influence the structuration of the space of norms to the same extent. (Actually, the space of norms contains a few norms that are not derived from the dominant ideology and thus there exists a variety of overlapping political cultures, but the fact remains that some norms are more pervasive than others.)

Now the methodological implications of the theoretical scaffolding mentioned above is that empirical studies of political cultures (i.e., of political "orientations") have very little predictive power.[15] The reason is that such measurements reveal nothing about the dynamic transformation of values into norms and of the feedback effects of the latter on the former. "Orientations" towards political objects reflect the transient equilibrium between conflictual values which prevails at one particular point. The more stable political values appear to be, the more likely they are to change, for in the space of values the dominant ideology periodically recovers critical, marginal counter-cultures when structural disorders reach a given threshold (the level of that threshold is determined by the existing norms). Consequently, new norms become acceptable after a usually rather brief period of transition (e.g., the transition from anti-statism to technocracy in Quebec in the early 1960s; conceptions about racial equality in the United States in the mid-1960s).

A theoretical grounding for this argument can be found also in A. Touraine's distinction between "historicity," whereby a society works upon itself, and, on the other hand, organizational functioning, i.e., the self-regulation of concrete societal structures; the former roughly corresponds to my space of values, the latter to my space norms, and between those two Touraine posits the existence of a political system where destructuring/restructuring choices are made and to which my prism of politics is analogous.[16] Measurements of political orientations that are available in the literature are taken in the fields of organizational performance and possibly of political decisions but, unfortunately, never in the field of historicity. At the same time, an exclusive preoccupation with ideological dynamics risks to degenerate into utopia if not logomachy.

In brief, my methodological prescription calls for a parallel and open-ended exploration of archetypes, values and norms. The result will most likely consist of a series of uncoordinated and possibly inconsistent observations. This is the cost of complexity and it is preferrable to the reductionist certainties offered by more traditional approaches.

From the Particular to the General and from the General to the Particular

The second paradoxical problem is that of the relations between the micro and macro-levels of analysis. A political culture does not exist independently of the individuals who are not only its vehicles but also its makers—that is, the whole is less than the sum of its parts insofar as the reproduction of the former is conditioned on the autonomy of the latter. On the other hand, an individual's personality is only an imperfect reflection of the cultural context that nurtured it—the whole is more than the sum of its parts since the very existence of the parts is predicated on the emergence of holistic qualities.

When studying political culture, more consideration ought to be given to the micro-level. This is imperative in spite of the fact that most research in this field is supposedly taking place at the micro-level because of the behavioralists' bias in favor of methodological individualism. Averaging out individual answers to standard questionnaires about political orientations "controlling for" a few conventional socio-economic variables creates an impression of (statistical) order where interesting contradictions may warrant the analyst's foremost attention. By contrast, to posit the existence of an autonomous collective being reproduced by means of a cultural code is a step towards the investigation of a wider range of attitudes and of more differentiated levels of recursive integration. (I have already dealt with the first aspect of this challenge in the context of the previous discussion of the impact of the terms norm, value, and archetype.)

In combination with the three phenomenological levels of meaning (i.e., experience, consciousness, and self-consciousness), at least three fundamental levels of recursive integration must be taken into account: the individual, the group/class and the polity. Two caveats are in order here. First, I do not claim to have discovered anything new: there exists a large body of literature on the effects of the levels of analysis.[17] But I would like to suggest that the substantive problem posed by the *autonomy* of various levels of integration—as opposed to the technical (logical) problem of inference from one level of analysis to the next—has not been dealt with properly. The way in which I intend to approach this problem also takes care of the second caveat, to the effect that in making the case for a broader perspective I do not mean to suggest that each and every empirical study of political culture ought to be equally concerned with the three fundamental levels of analysis, but rather with their paradoxical overlapping.

Each one of these three levels of integration can be subdivided in a variety of ways. From a bio-cybernetic point of view, the criteria to be used in this respect are that i) each subordinate level should contribute to the autonomy of the level of integration considered, and ii) at least one of these subordinate levels should be defined in such a way that it could be related to the next level of integration. For example, individual political actors can be interesting for their presumed representativeness (e.g., a sample selected randomly in a given population) or, on the contrary, for their atypical character. The latter phrase refers to individuals whose orientations and values, although probably not very "representative," may help the observer understand the concrete ways in which autonomous political actors take part in the construction of an autonomous public realm (e.g., social and political leaders, militants, intellectuals, etc.). Thus the evidence obtained from surveys should be confronted with analyses of the life-worlds of non-randomly selected individuals (the techniques to be used in this respect range from psychoanalysis to historical biographies).

Similarly, in the study of ethnic, religious, etc. groups or classes, provisions should be made for the investigation of panels representative of the various "elites" or vanguard movements that are active within the group or class considered. While in a sense being the active "conscience" of that group or class (e.g., trade union militants vis-a-vis the working class at large), these segments may also find themselves in a position to transcend their particular ideologies—would it only be in order to adapt them to ecological variations. An example of this kind of ambiguity which unfortunately cannot be discussed at length here is the question of deciding to which class the state bureaucratic elites belong. Of course, they belong to the dominant class "in the final analysis," but their autonomy has become so extensive in most developed countries as to render such a characterization almost meaningless. In any case, more comparative studies are needed.[18]

As far as the third fundamental level is concerned, it is crucial not to reify national myths by supposing that they are shared unanimously or with the same degree of intensity throughout the polity. Regional political cultures often constitute antithetical poles. (Note also that regions may be connected with other nation-states thus opening up the polity to the international scene from within, e.g., Alsace vis-a-vis Germany; this is even clearer in the case of multinational states, e.g., Lithuania vis-a-vis Poland,

Quebec vis-a-vis France, etc.). This problematic political unity in diversity is, for example, characteristic of Canada.

The State as the Organizing Nexus of the Polity

If, in some sense, national myths embody the identity of political communities, as well as that of the individual members of these communities,[19] they do not concretely produce that community. So far, I have attempted to circumscribe the "self" of a self-organizing polity but I still have to specify the organizing principle at work in the latter. The fundamental difference between the concepts of structure and of organization has already been explained;[20] its relevance will become evident in view of the fundamental questions that it suggests in the present context. I will argue that the state can be conceptualized as the organizing nexus in a given political community. Further, I will show that this perspective helps us to accommodate a variety of theoretical positions about the nature of the state. Finally, I will venture that an organizationalist paradigm promises to make possible the reappropriation of the major themes of classical political theory by a more empirically oriented type of political inquiry.

Of course, the state, under the form of what is usually called "government," is also a structure interacting with other societal structures. I am interested here in the state as an unstable form of societal association and I will look at it from both an empirical and a semiological point of view. (From that latter angle, the state appears as the symbolic space where cultural values are transformed into ordered norms of action, linguistic policies are a case in point.) However, it is important to realize that this definition may also imply that the state is a stable, homogeneous structure; in fact, it is neither. The state is composed of individual political actors who retain their autonomy.

The ordering intervention of the state is something that has been recognized by many authors. For example, W.J. Stankiewicz explains that, until the late modern times, it was thought that the role of the state was to order norms.[21] The contemporary triumph of relativism, according to him, has obscured this fact. And the effects of such a pernicious refusal to acknowledge the obvious, so to speak, are felt at the analytical as well as at the practical levels. Behavioralism reflects an incapacity to conceive norms as something else than the sanctioning of transient dispositions; thus Stankiewicz contends that the behavioralists' world-view looks very much like Hobbes' state of nature. But, while Hobbes used the idea of a primitive state of nature simply as a means for discovering the origins of politics, behavioralism cannot look beyond it. (Implicitly, Stankiewicz treats behavioralism as a pre-political vision: in their refusal to take time seriously, the behavioralists have locked themselves in a mythical dark age!) From a practical point of view, Stankiewicz notes that

> The central norms of our society now derive from the State rather than the Church, but the State does not have any system of ordering norms; it pretends that the family and Church will serve that function. The point has now been reached where the State must begin to reappraise its position. It must be remembered that norms are still with us, but—being 'unordered'—they contribute to our aimless behavior. We are relativists only in the sense that our norms are disordered.[22]

On the other hand, Etzioni offers us a formulation of the same problem which is more in keeping with the language of the social sciences but which is also less probing and therefore less suggestive. For him, the state is "the tool of societal mobilization and action." The organizing role of the state is understood by him in relation to "consensus building"; that is, "Societal consensus building in post-modern societies . . . is focused around the use to be made of the state and the way it is molded and remolded."[23] The kind of consensus to which Etzioni refers here is an agreement over goals and priorities and, therefore, his perspective converges with both Stankiewicz' and my own views on the role of the state in the space of collective representations. However, Etzioni's refusal to acknowledge the normative implications of his position limits the import of his approach. These references support the idea that the ordering intervention of the state is a legitimate premise in social and political theory. It remains now to see how this premise could help us to sort our way through the contemporary debate on the role of the state.

Each in their own way, behavioralism and Marxism have contributed to the evacuation of the state from the domain of political inquiry, at a time when the activities of the state have expanded tremendously in all industrialized countries. The present crisis of the welfare state, however, has prompted a reappraisal of these conceptions. Indeed, the problem of the role of the state now stands at the top of the political agenda at both the right and the left ends of the ideological spectrum.

Grounded in both liberalism and pluralism, empirical social science has been characteristically "society-centered," to use Eric Nordlinger's term.[24] That is, the democratic state is supposed to be subjected to the determining influence of a variety of societal constraints. For example, the idea that socio-economic factors like the GNP per capita or the rate of urbanization constitute the explanatory variables that account for the greatest proportion of the variance in levels of spending on comparable programs among several countries (or among the fifty states in the United States) is an assumption that is common to a whole range of comparative policy studies.[25] This is entirely consistent with the deterministic and reductionist bias inherent in much of contemporary empirical research. Such a lack of concern for the values and preferences that state regulators have acquired makes it very difficult to comprehend the dynamics inherent in societal self-organization for it ignores some of the most significant sources of socio-political "noise." At the opposite end of the spectrum, E. Nordlinger's "state-centered" model is predicated on the assumption that state elites often have the ability to translate their own preferences into authoritative action even when state and societal preferences are divergent. They can either implement policies that challenge the interests of some powerful groups or they can at least "purposefully bring about a shift in societal preferences to make them congruent or consonant with their own, followed by the translation of their now non-divergent preferences into public policy."[26] I concur with Nordlinger when he stresses that the state is not to be conceived as a single entity but rather as a network of decision-makers whose individual preferences matter. For if the nation is a collective being, the state is not; it is merely the insitutional setting within which identifiable public officials pursue goals that may or may not be congruent with those of the most powerful societal groups. The mythical unity of the national community (i.e., expressed through meaningful myths) and the historical reproduction of something like a "national being" do not entail the absence of conflict in the polity. The problematic and always unstable congruence of state and societal preferences is a fundamental aspect of political life; all political ideas, structures and processes depend on this uncertain relationship. Nordlinger's "state-centered" model is not only more plausible, as he puts it, than the society-centered one which has proven to be so influential, but it is also a useful methodological contribution to the study of the self-organizing polity.

For orthodox Marxism, the state is an instrument of the ruling class; it is no more than "a committee for managing the common affairs of the whole bourgeoisie."[27] Neo-Marxist authors like Ralph Miliband, Nicos Poulantzas and James O'Connor have reacted against the simplistic economism of the orthodox view.[28] They argue that this bias masked the very real powers exercised by a relatively autonomous state apparatus. At the very least, it must be assumed that the "common affairs of the *whole* bourgeoisie" do not correspond necessarily with the short-term interests of all its fractions. On the basis of Althusser's concept of over determination, Poulantzas maintains that the state is actively engaged in the politics of class struggle rather than forming merely a passive reflection of the basic contradictions inherent in the capitalist mode of production.[29] For O'Connor, not only does the state enact policies that foster capital accumulation, but it also uses coercion to implement policies that foster social harmony (e.g., social security programs); in so doing, the state may work against the interests of most of the fractions of the bourgeoisie, although in the long run it helps to legitimize the capitalist system as a whole.[30] This, however, is a moot point: the extent to which the autonomy of the modern state is "relative" is a matter that cannot easily be determined a priori. While it seems clear that in many third world countries, especially in Latin America, the autonomy of the state is very relative indeed, the complexity of state-society relations in advanced industrial societies renders this kind of classification rather meaningless.

If each in their own way empirical political science and neo-Marxism have rediscovered the importance of the regulatory functions of the state, these have ceased to be simply an empirical problem and have become the object of a normative debate prompted by the political successes realized by neo-conservative forces in much of the western world. Under these conditions, it seems necessary to reflect anew on the place of the state in the polity. I begin this reflection by defining certain key aspects of the generic notion of the state. It will be argued further on that the state plays an indispensable function from an evolutionary point of view, a function that can only be comprehended by means of a resolutely normative inquiry. A reflection on the practical meaning of separating public from private activities is offered as a means of escaping from the narrow field on which the battle between supporters of the status quo and the neo-conservative forces is raging.

The Status Rei Publicae

What is the nature of the societal complex we call the state? The term itself derives from the Latin *status*, as in *status rei publicae*, but its use dates back only to the 17th century. (The Romans spoke instead of the *res publica* and the Greeks of the *polis*.) The *status* invoked here is that of sovereignty which guarantees the wholeness of the polity. Today, the sovereignty of the state is more ambiguous than that of the monarch was for Hobbes, but it remains an important symbol.[31]

Political behavior has no meaning unless it is directed toward the production of some identifiable connection in an existing or emerging societal structure (e.g., voting works toward the production of a link between "doers" and "done-by"). The purpose of political action is to enhance the capability of the polity to act upon its environment or upon itself in an autonomous manner. Thus one would waste time studying voting behavior in the Soviet Union or in any other one-party régime, precisely because in such a case the state is not sovereign—only the party is.[32] Traditional political theory acknowledged that state sovereignty is a *sine qua non* of political life. This idea has been expressed in a variety of ways (e.g., Jean Bodin, Thomas Hobbes, the 19th century legal positivists, etc.). While many classical thinkers can be faulted for going too far in extolling the virtues of absolute sovereignty, most modern political analysists err in reducing sovereignty to the exercise of raw power.[33]

To put it in terms consistent with a cybernetic approach, politics supposes not only "energy" (i.e., power) but also "information" coded by a system of norms which is the product of evolutionary learning, as we have already seen. Recalling earlier developments, learning entails closure and consequently autonomy is a central concept in cybernetics. A neo-cybernetic approach, because it further posits that complex systems possess a self-created identity, allows us to transform by analogy the cybernetic notion of systemic autonomy into that of state sovereignty, while retaining the suggestive ambiguity inherent in the concept of control. A controlled system is never fully closed to its environment and furthermore, a learning system, is open to the "structuring information" (Laborit) that permits it to move to a higher level of complexity. Thus our analogical understanding of sovereignty also implies a rejection of statism and invites us to reflect critically upon the manner in which the state can realize its own diversity.

To continue our exegesis of the phrase *status rei publicae*, we now have to ask: Whose public concerns? The state is a collectivity which is usually, but not necessarily, cosubstantial, although never coinciding, with the nation. Some nations either antedate the state (e.g., Germany, Italy), or have existed for long periods of time without a state (e.g., Poland), or form only part of a state (e.g., Quebec). On the other hand, many of the states that have been created during the last few decades rule over yet amorphous nations. There is no need to elaborate upon this point which is discussed at great length in the literature on political development. The nation-state is not therefore *the* only conceivable polity. But it seems clear that in stateless nations (e.g., Poland in the 19th century) politics remains outwardly, being either a ritualistic revival of liveless myths or an eschatological projection into a promising future (this may help to explain the political influence of the Catholic Church in Poland, or in 19th century Quebec).

Nationless states constitute self-regulating structures whose performance produces unmistakenly political communications. But are they also self-organizing systems? Or what degree of control over the morphology—as opposed to the physiology—of political life do the members of such states enjoy? I submit that the answer is: very little. As long as there can be no sense of common purpose revealing the unexploited potentials generated by societal "noise," it seems highly unlikely that strictly functional political systems could do anything else than replicate themselves identically. If so, the phrase "political development" is something of a misnomer in the case of new nations (especially in the case of African "nations").

The term *publicae* leads one to inquire: What affairs are public or what does the state produce? At the level of analysis adopted here, i.e., the conscious representations of the unity of the polity, the state takes the form of a legal system. But this is a static qualification corresponding to a static approach. In a dynamic perspective, the state converts power into authority. (This does not mean, of course, that members of the polity never experience the coercive force of the state.) From Max Weber to Habermas, legitimation has been thoroughly debated, and there is no need to restate here all the stages of this debate. But I wish to address a question which lately has become topical again: Is there a case for minimizing the role of government?

Whither the State?

Having delineated the concept of state and established its import for the conscious representation of the polity by its members, we can understand better now the shock that the emergence of the modern state must have caused in the minds of early 19th century thinkers. With hindsight, we should be able to avoid falling into the trap of a statist ideology, but we must also remain critical of the contemporary topsy-turvy neo-Hegelianism according to which the state incarnates all (historical) evils.

To begin our reflection on the state, we must ask two questions: Is the state a dispensable historical artifact and, if this is not the case, how can the state be used for the realization of a humanized world, for reaching the "good life"? The theory of self-organization adduces at least metaphorical reasons for answering negatively to the first question. I argue that a viable state is no more, but no less, than a prerequisite for the practice of a creative politics. However, before taking up the question of knowing what should be expected from the state, let us first consider the problem of the dispensability of the state.

The theory of self-organizing systems is evidently not a theory of the state, but it may help us to lay the ground-work for some novel theorizing on the production of political order. Let us recall the logical interdependence that it establishes between order and disorder. More specifically, it has already been argued that what is invariant or changing in a given system depends on the axiomatic intervention of the observer. It is only the invariance of some organizing principle that renders change meaningful (e.g., mutuations derive their significance from the presumed invariance of the genetic code). In moderately complex systems, the invariant coordinates remain implicit. However, in truly complex systems, these coordinates need to be clearly identified. Thus, for example, biologists will look at the organization of the gene pool or, better still, at the biological identity of the phenotype in their search for the coordinates of biological evolution. Note that these invariant coordinates are not absolutely invariant but simply more stable than the processes that they organize. Ultimately, the latter reproduce and transform the not so invariant organization itself. If it could be shown that the state represents such an invariant organization at the stage of societal evolution in which we find ourselves at present, the indispensability of the state would thereby be established.

This problem can be restated by means of a distinction between power and "meta-power." As it is probably the case even in the most stable societal systems (presumably "primitive" ones), and certainly in developed ones:

> Human actors deliberately try to alter the existing matrix of action possiblities, outcomes, and orientations within which social action occurs. We [i.e., Baumgartner *et al.*] refer to the exercise of such "meta-power" as *relational control*, that is control over social relationships and structures, which we distinguish from power within a given structure: clearly, although an actor may have social power within an interaction framework or "game" . . . he may not have power to structure social relationships, to alter the "type of game" the actors play, or to manipulate or change the distribution of resources or the conditions governing interaction or exchange among the actors involved.[34]

While in pre-modern societies the rate and scope of politically induced change is such that meta-power remains somewhat diffuse, it seems reasonable to hypothesize that an acceleration of societal change requires the conscious apprehension and explicit legitimization of meta-power. Then the body politic becomes a state. (Note that by tracing the origin of the state to the acceleration of societal change, a *deliberately* vague phrase, I implicitly join the ranks of those who, like P. Corning, E. Service, or G. Geiger argue that no single factor can explain the emergence of state institutions; examples of monocausal explanations, on the other hand, would include K. Wittfogel's emphasis on the role of irrigation systems in the development of early civilization, or H. Spencer's preoccupation with external conquest.)[35]

Jean-William Lapierre defends a rather similar thesis.[36] He explains that in animal societies collective behaviors are regulated by means of more or less flexible, more or less instinctual relations of domination/submission. But animals never enter into relations involving command and obedience, for these presuppose the recognition of a legitimate authority. Not only do human societies differ from animal societies in that they practice relations of authority, but the forms taken by these relations vary greatly among human groups. Here Lapierre takes issue with anthropologists like E.E. Evans-Pritchard or P. Clastres who contrast societies with and without a state. On the contrary, he posits the existence of a continuum in seven stages extending from societies that are indeed completely stateless, i.e., where political power is diffuse (e.g., the traditional Inuit society), to modern bureaucraticized societies.[37]

From his review of much of the political anthropology literature covering one hundred traditional and historical societies on all continents, Lapierre concludes that there is a definite, albeit not rigorous, correlation between the degree of political organization and the adaptability of a political community to internal and external aggressions.[38] To put it differently, societies that are unable to harness meta-power cannot withstand major disorders; all they can do is to respond homeostatically to small variations away from routine patterns. The history of the European colonization in Africa and in the Americas clearly illustrate the fragility of societal systems lacking the instruments of statehood.[39] Most of them were incapable of legitimizing social innovation.

The point here is not, of course, that Western civilization is inherently superior to all other forms of socio-political life. It is rather that there are significant trade-offs to be made between, on the one hand, the intensity and scope of coercion and, on the other hand, the immediacy of the source of legitimacy and the freedom to innovate. Members of stateless societies are remarkably free and equal in their interpersonal relations (at least among members of the same sex), but they must submit themselves without reservation to a source of legitimacy which is not only external but also somewhat hidden (i.e., it cannot be fully comprehended and thereby transformed). This deprives them of the freedom to choose alternative courses of individual and/or collective action in order to face the challenges of a changing environment. Lapierre's account of the evolutionary potential of statehood echoes Habermas' hypothesis that the formation of a state is the second of four evolutionary stages (although Lapierre's approach is more empirical).[40]

Modern politics is the field of human action where contingent historical events actualize evolutionary potentials into concrete societal structures, and Lapierre devotes almost as much attention to historical as to anthropological data. Taking the English revolutionary era as his example (i.e., the chaotic transition from a traditional to a parliamentary monarchy in England between 1640 and 1688), he shows how political revolutions are the culmination of a series of "social innovations" (e.g., the formation of new classes of bourgeois and small merchants in the cities, of squires and prosperous yeomen in the country); and how the cumulative effects of these innovations could not have been secured without political change.[41] The state enables a political community to innovate. This is not to say that politics is not an autonomous sphere—although Lapierre, who often attacks Marxist theory for its reductionist bias, is here strangely ambiguous. It is rather to point out that the complexity of a society's political organization is both a necessary condition, and the outcome, of the society's ability to transform itself. To take Lapierre's argument a little further than he may be prepared to pursue it, one could see here a clear instance of political self-organization: the legitimation of new societal structures, whose emergence is itself conditional upon the relative openness to change of the legitimation process, brings about a qualitative transformation in that process.

We owe to A. Khoshkish an interesting reformulation of the three Weberian ideal types of authority in a manner that evokes the premises of the theory of self-organization.[42] For Khoshkish, the conversion of power into authority takes place within a triangular structure the three poles of which are: i) the "domineering," and ii) the "dominated" components of the "power complex," and iii) the legitimizing agency. Historical development consists in a progressive rationalization of the cultural processes through which the legitimizing agency is produced by the power complex. From an organizationalist point of view, this means organizing the production of structures of political interaction in such a way that these structures ultimately produce their own legitimacy, and thereby transform it. For Khoshkish, the transition to modernity corresponds to a change from the "consecration" to the "constitutionalization" of power. The former involves the recourse to an external source of legitimacy which is of a spiritual and/or traditional nature. The latter is

> . . . the process by which legitimization of power into authority develops mainly within and among the components of a power complex, with little or no intervention by or resort to an external factor—in particular supernatural.[43]

There are three dimensions to this process, i.e., traditional, contractual, and representational—the representational dimension, which subsumes the history of parliamentary democracy, being the most typical one. In view of Khoshkish's insistence on the conflicts that develop between the various dimensions of each mode of legitimation, especially between the spiritual and traditional dimensions of consecration (Khoskhish cites in this respect the conflict between the *sacerdotum* and the *imperium* in medieval Europe), one may be entitled to reinterpret his position as meaning that self-organization of the polity supposes a measure of semiological disorder. The ideal political order is certainly not an immutable concept. Even if the pace of organizational change is definitely slower than that of structural change, conflicting ideas develop and compete in the space of legitimacy. Plato's intriguing account of the "natural" fall of the ideal state still stands as one of the most original and provocative discussions of this subject.[44] Space is lacking here to review critically all that has been written since then on the same subject, but it is clear that the uncertain legitimacy of the modern state is a dominant theme in social and political theory today, at both ends of the ideological continuum.

For example, J. Habermas argues that the greatly expanded integrative functions of the state, whose planning activities create legitimation problems in spheres which used to be self-legitimized (e.g., cultural activities),[45] is not accompanied by the development of democratic political participation. On the contrary, the state resorts to technocratic practices that bypass politics. For Habermas, though, only a political discourse can generate the kind of meaning which is lacking in the system. Thus, the risk of an authoritarian drift appears to be very high.

On the other hand, several authors have diagnosed a growing impotency of the state caused, in their views, by a rise in the power of interest groups that have succeeded in building up a local or sectoral legitimacy. In the absence of a societal consensus, it becomes more and more difficult for the state to carry out its coordinating role. This is otherwise known as the "overload" thesis.[46]

Why are contemporary democratic states overloaded with demands is a question that the by now popular "public choice" theory explains in terms of the cumulative effects of the interest-maximizing strategies pursued by the voters, the politicians and the bureaucrats.[47] The politicians typically seek to maximize their electoral support, and the voters tend to vote for the politicians whom they believe are ready to implement policies that will benefit them. (Voters estimate the costs of a policy in terms of taxation rates, and the benefits in terms of the value of the services or subsidies that they receive from government.) As for the bureaucrats, they are interested in increasing the size of their agencies' budgets or, more generally, in strengthening their power within the organization. Because taxpayers who do not benefit from a given program are not likely to oppose its implementation, since the cost of doing so would outweigh the cost of the proportionally small increase in taxation which is due directly to that program, whereas those who benefit from the program are much more likely to be active and vocal supporters of its implementation, it follows that politicians find it more advantageous to expand rather than to reduce the scope of government intervention in society. Such a paradigm explains some of the growth of the public sector between the end of the Second World War and the late 1970s. Similarly, the politics of restraint which prevails today in North America and in much of Europe could also be explained in terms of the unpopularity of the bureaucrats whose position is often terminated in the name of deficit cutting, and in terms of a revolt of the middle-class against taxation upon which politicians capitalize.[48] In brief, the public choice school reduces the state to a collection of agents who seek to maximize their self-interests by engaging in various transactions in a political market that functions very much like a market economy. It describes political life as being essentially similar, if not altogether subordinate to economic life. Therefore, the public choice paradigm belongs to the broader universe of neo-conservative theories which attempt to repudiate the notion of the positive state. (Actually, the label "neo-liberal" would be more appropriate considering that neo-conservatism is in many respects a return to classical liberalism;[49] unfortunately, "liberal" today in North America means someone who is an advocate of the positive state.)

From fighting rearguard battles against Keynesian economic management and against the growth of the welfare state in the 1960s, the neo-conservative forces have moved into a dominating position both intellectually and politically in the 1980s. Most of the issues on the neo-conservative agenda are economic in nature; monetarism, laissez-faire, and free-trade are the fundamental tenets of the new orthodoxy. The state, from this perspective, is to play a reduced role because it is argued that i) government spending on goods and services reduces capital investment and inhibits economic growth; ii) inequality reduction through taxes and transfers reduces output and growth; iii) minimal taxation of business is necessary to attract foreign investment; iv) government bureaucracy is inefficient and wasteful; and v) government spending does not provide "real" jobs.[50] Whereas much of the literature on policy-making in the 1960s was concerned with the achievement of macro-economic goals, the emphasis today is clearly more on micro-economic efficiency.

However, the modern version of the minimal state theory goes beyond the confines of economics. It is the foundation of a political philosophy which has been articulated by various thinkers, none more eloquent perhaps than Friedrich Hayek. The welfare state is dangerous, according to Hayek, because it violates the Rule of Law. And the Rule of Law is our best protection against tyranny, insofar as it is a "meta-legal" principle that sets limits to the scope of the laws enacted by the legislative branch. Unfortunately, modern legislation takes the form of instructions to civil servants rather than being a legal definition of the relation that should prevail between the citizens and the state. Therefore, the welfare state will use methods that are often inimical to liberty:

> If government wants not merely to facilitate the attainment of certain standards by the individuals but to make certain that everybody attains them it can do so only by depriving individuals of any choice in the matter. Thus the welfare state becomes a household state in which a paternalistic power controls most of the income of the community and allocates it to individuals in the forms and quantities which it thinks they need or deserve.[51]

More recently, "neo-conservative" authors like Irving Kristoll or Daniel Bell have pursued this analysis one step further. Their critiques of the politics of redistribution has led them to challenge even some of the tenets of pure liberalism, in particular what they regard as its spiritual emptiness.[52]

Centralized state planning, social engineering and the generalization of bureaucratic controls are tendencies that are hardly consistent with the idea of a self-organizing polity. To that extent, I share with the neo-conservative critics a concern about the excesses of the positive state. However, the self-organization paradigm points to several weaknesses in the neo-conservative position. First, the size of the state is a very imperfect indicator of the degree of political freedom enjoyed by the citizens of any state. What really counts is the ability of the political community to steer away from the path or chreod that leads to stagnation and decay. The state is certainly not the only means of adapting and restructuring societal institutions and interpersonal relations, but, in some cases, it is incomparably more effective than any other one. As new norms become accepted, the state, especially the democratic state, can use an array of policy instruments, ranging from consultation to coercion, which no other societal structure can match in order to make the new norms effective. For example, women and minority groups have achieved gains in North America and Western Europe which would not have been realized, despite a more tolerant ideological climate, without state regulation. Paradoxically, the transition from the positive state to the minimal state would involve such profound societal changes (and might provoke such powerful reactions) that, at least in the short term, it would require a strengthening of the existing state institutions and the deliberate exercise of coercion.[53]

Secondly, the liberal insistence on the primacy of the principle of individual freedom is entirely consistent with the neo-cybernetic paradigm which rests on the assumption that autopoietic unities retain their autonomy when they interact with each other. However, the singular emphasis on competitive behaviors which is inherent in the economic and political models that we associate with neo-conservatism implies that individuals are not free to engage in any form of social relation that is not competitive, lest it result in an inefficient allocation of scarce resources. It may well be that market economies are most efficient when competitiveness is least constrained, but this hardly does justice to the complexity (i.e., multi-dimensionality) of societal relation, and of political relations in particular. Economic determinism is the Achille's heel of, not only Marxism, but also of classical liberalism. From a neo-cybernetic standpoint, however, the economy is just one of the structures that compose a self-organizing polity.

Thirdly, and most importantly it is clear that a market economy is a self-regulating system which, in most instances, is more efficient than administered economies. But it is *not* a self-organizing system. That is, a market economy is reproduced by the polity as a whole. The norms that regulate exchanges of goods and services on national and international markets are often authoritatively determined by political means, or evolved through customs, religious practices, etc. They are not spontaneously generated by the economic institutions in isolation from the other societal institutions. The continuing existence of a market economy hinges on the continuing existence of a legal system, of courts, and of other government agencies, and the survival of the private sector may, at times, require a strengthening of the public sector (e.g., in most countries, exporters rely on government assistance in the form of loans, subsidies, or marketing schemes).

In brief, we have seen so far that there are conflicting ways of theorizing about the state. The approach proposed here does not make it possible to settle these differences once and for all. It offers, nevertheless, the opportunity to take a new look at these various positions. It posits that the part played by state authorities in the reproduction of the polity as opposed to their strictly regulatory functions, is a theoretical problem in itself. The organizing role of the state is not contingent upon any particular set of historical circumstances; consequently, the study of politics is, to a very large extent, an inquiry into the course of societal evolution. In particular, a neo-cybernetic approach forces the analyst to specify what, if any, societal practices are emerging that challenge the organizing role of the state. Could these practices evolve into an alternative mode of societal organization? Could they produce their own, qualitatively distinct, legitimacy? Unless these questions are answered, it looks as if the difficulties encountered by public authorities in Western democracies when they attempt to legitimize their multiple interventions in almost all aspects of ordinary life reflects perhaps less a crisis of the state as a whole, requiring an immediate and radical change of values, than an imbalance within the executive branch. This imbalance results from the depoliticization of government, that is to say from the uncritical adoption of a centralized and overextended mode of technocratic policy-making. But the democratic state as a whole may still be healthy. The various forms of resistance that bureaucracies meet today signify that the state, as a political community, is not in crisis, but rather that it has produced the crisis. The uncertain triumph of technocratic rationality, i.e., the reduction of reason to instrumental calculus, and of politics to administration, does not result from some irresistible societal determination but from an inaccurate reading by some members of the state elite of the current, and changing, perceptions of what constitutes the common good. The most important political problem today is not simply that of redefining the role of the state; it concerns the scope and organization of the public realm, a phrase that refers not only to the state but also to the symbols, values and norms contributing to the reproduction of the political community.

The Recovery of the Public Realm

As we have just seen, a neo-cybernetic approach to politics leads to the analysis of normative questions. What follows, however, is only meant to be an exploratory study of the substantial questions that a thorough analysis of the place of values in political life would imply.

To bring classical authors back in focus, as Leo Strauss and his followers have done, does not *ipso facto* establish the relevance of classical political theory. It remains to be seen why, and to what extent, the questions that preoccupied the great thinkers of earlier ages still concern our contemporaries and, furthermore, that their methods have not been superseded by more powerful modes of reasoning.

The political vision that best characterizes the anthropological discourse of the new cybernetics portrays man as a political animal, insofar as his sense of self reaches to the level of a self-organizing public sphere. There, meaningful values are created through complex interactions that cannot be accounted for entirely by socially induced "desires" or by objectively determined socio-economic variables. Although such a turn of phrase does not compare with the simple elegance of Platonic dialogues, it may evoke the Ancients' understanding of man's bond to the polis.

Now, there is little to expect from an impossible return to the world-view of ancient times.[54] But we can benefit perhaps from a searching and critical continuation of Hannah Arendt's untimely reinterpretation of classical thought, and especially from her rediscovery of the "public realm."[55] The new cybernetics adduces a few reasons for proclaiming that the time has indeed come for a reinvestment of the public realm, and for a renewed interest in political action.

The cyclical movement from closed to open states in self-organizing systems parallels, in a formal manner, Arendt's criticism of the closure of the modern world at the level of the private enjoyment of the fruits of one's work. Actually, Arendt's pessimism is overstated because the increasing complexity of environmental and socio-economic problems has revealed the limitations of the private sphere. But the technocratic response to this situation, i.e., the ill-fated attempts to implement one kind after another of centralized planning, has turned out to be the source of even more uncertainty. Most of these planning models rest on a premature cyberneticization of societal interactions. Whereas the first cybernetics ignored the question of the origin of norms, the new cybernetics poses it as a prerequisite to the investigation of any kind of oriented process. From a political point of view, this task looks very much like a recovery of the public realm: the production and reproduction of societal order supposes that societal actors, by publicly articulating their respective wants, reach a (shadowy) consciousness of their unsatiability. This consciousness helps to transform these wants into norms of purposeful and publicly legitimated action. In other words, a norm consists in a projection (onto both a time scale and an intersubjective space) of tensions that cannot be solved immediately and individually. Among these multiple tensions, the various ways in which we relate to state institutions are the most relevant to this inquiry about the nature of the public realm. (I have alluded already to the less conflictual and more symbiotic relations that exist between our private selves and the collective self which is formed when we share with others a common cultural/national identity.)

The liberal idea that individuals find in themselves alone the source of their uniqueness is the cause of the modern fascination for the inner world of consciousness; but it has led also, to a privacy fetishism and to the neglect of the public realm. Perhaps the most obvious manifestation of this inability to think of political action as a meaningful way of life can be found in the spontaneous association that we commonly make between the state and bureaucracy—a conceptual linkage which is consistent with experience, it must be conceded. The gap between the theory and practice of modern democratic government confronts the political subject with a distorted image of himself as a political actor, on the one hand, and as the powerless client of bureaucratic organizations staffed by faceless bureaucrats, on the other hand, forcing him to retreat into an inner sanctuary. Viewed from outside, the public sector is sometimes benevolent, even beneficient,[56] but too often it evokes a sense of Kafkaeske absurdity. Viewed from inside, as Ralph Hummel suggests, the bureaucratic state is a world altogether different from the rest of society; in particular,

> The self-concept of the functionary in bureaucracy . . . is never allowed to develop in terms of separateness and uniqueness, but only in terms of integration and similarity (or functionality) in relation to the rest of the organization. What the individual functionary is the organization intends to design exclusively from above.[57]

(Note that a criticism of the idea of the self "in terms of separateness" can be read in my previous attack on the liberal concept of the self; but I still think that Hummel's statement is relevant insofar as I do not propose to resolve the tension between the individuality of the bureaucrat and the hierarchical principle by arguing that private norms ought to be replaced arbitrarily with official ones.)

It would be senseless to pretend that these tensions can be solved once and for all, and, in any case, they can only be worked out in a practical—as opposed to an analytical—manner. Therefore, what follows is proposed only as a means of coming to terms with these problems in a way that opens up a new perspective.

A common sensical answer to the question of deciding what defines the public realm would consist in saying that it has something to do with the public interest. And indeed it has: state activities that do not result in the achievement of some goal or societal condition generally regarded as desirable would not be legitimate. At least this aspect of the contractual theory of political life has become tautological. But the public interest—admittedly, an already reductionist interpretation of the notion of the common good—

is a rather meaningless concept. In practice, the public interest is often given an ideological definition. Some, for example, will argue that the public interest is a convenient façade for the interests of the ruling class or for the selfish goals of the state elites, while others believe that the public interest is coterminous with the organizational requirements of a benevolent welfare state or, on the contrary, with the functioning of a market economy and with the pursuit of a laissez-faire policy. This is not to say that it is futile to attempt to reflect upon the practical significance of the common good. Too often, for example, policy decisions are based on the result of cost-benefit analyses that are predicated on utilitarian premises which ought to be the object of a thorough reexamination by political theorists and ordinary political actors concerned with the ethical implications of public policy. However, political problems demand practical rather than principled solutions. The domain of action is open to agents who share, or fight about, not only ideas but also experiences. Thus the minimal state and a deregulated economy are bound to be somewhat less "minimal" or "deregulated" in countries that have developed lasting administrative traditions (e.g., China, if it ever goes that far, or more appropriately France or Germany) than in traditionally liberal countries (e.g, the United States).

My contention is that a self-organizing national being forms the context of reference in relation to which one can pose the twin problems of the organizational autonomy of the political system and of the scope of the public realm. It is not sufficient to distinguish private from public concerns, or consenting from coerced interactions, it is also necessary to distinguish those activities that contribute to the (re)production of a national identity from those that do not. For example, the on-going debate over the comparative merits of laissez-faire liberalism and of state interventionism is not likely to be settled conclusively until one recognizes that the valuation of any system of resource allocation depends not only on the relative efficiency of this system, but also on many other criteria relevant to the individual and collective experience of the political actors who are in a position to influence this process. Thus, if the practice of economic liberalism is associated with a very unequal distribution of wealth and with wasteful consumption, as is often the case in Latin America, only those who directly benefit from the status quo will be opposed to coercive interventionist measures such as land reform. On the contrary, the reproduction of the American polity supposes the reproduction of a healthy market economy and of many "negative" freedoms. The most successful reformers are those who are able to convince the American public that their vision is consistent with the liberal values inherent in the American political culture: Franklin D. Roosevelt did, but most Democratic presidential candidates failed to do so since the late 1960s. Yet even in the United States it is recognized by most that poverty and unemployment are matters that can tarnish the image that

Americans hold of their own society and are therefore legitimate public concerns that government must, at the very least, acknowledge and address.

Cultural (i.e., literary or artistic) life and fundamental scientific research provide us with an even more ambiguous and interesting case. The pursuit of artistic or scientific excellence and the enjoyment of the benefits that flow from such activities are either strictly personal experiences or experiences that acquire a universal meaning, but, in any case, not matters that are relevant only to the concerns of a particular national group. Regulation of these activities by the state is often odious: censorship, ideological repression (e.g., Lysenkoism in the Soviet Union, or "soviet realism") and manipulative distortions (e.g., the propaganda advantages derived from the national origin of men or women of genius) are practices that immediately come to mind in this respect. However, it is equally true that cultural messages play a decisive role in the production and reproduction of a national identity. Hence, the artistic and scientific communities are often engaged in quite legitimate transactions with their respective national governments and, in some cases, financial support from government sources and protectionist policies may be the only feasible means of preserving an original cultural tradition (e.g., the state of broadcasting in Canada). The practical solution to these dialectical tensions can only be worked out in a public forum.

In conclusion, the proper role of the state and, in particular, of the individual leaders whose values and preferences influence the direction of public policy is to cross-examine the national community, as it were, in order to confront it with its own image, as they see it. This process of image production will be all the more effective if there exists an institutional setting that allows a dialogue to take place among state authorities, and between them and the individual and groups who recognize themselves in the images that are articulated for them and by them. This is the positive role that even a limited government must play if it is to remain legitimate—but which may lead it to appropriate for itself a larger and larger portion of the public realm.

Notes

1. See, for example, B.C. Shafer, *Faces of Nationalism* (New York: Harcourt Brace Jovanovich, 1972).

2. Unexpectedly perhaps, in the light of what was said above about his insistence upon the characterization of a nation as a living being, Morin himself writes: "One can and ought to conceive another type of social system [i.e., other than the nation] in which yet unexploited hypercomplex anthropological virtualities could develop . . . and we can conceive a society of nations that would be truly revolutionary in comparison with the present reality of international relations and would render it possible to unite the whole human species into a federation." "La nature de la société," p. 30, my translation.

3. Quoted by D. Cameron in *Nationalism, Self-Determination and the Quebec Question* (Toronto: Macmillan, 1974), p. 44.

4. Ibid., p. 74. However, there are exceptional circumstances in which consent may actually be withdrawn (e.g., Norway's separation from Sweden in 1905) or envisaged, either seriously (e.g., the referendum held in Quebec in 1980) or hypothetically (e.g., the constitution of the U.S.S.R. contains a clause pertaining to the voluntary secession of any federated republic).

5. Jaurès wrote in the *Revue de Paris* (1898): "To break nations is to overturn sources of light . . . It is to suppress centers of rapid action, leaving in their place the incoherent slowness of universal effort. Or rather, it is to suppress all liberty; because humanity, no longer in a position to condence its action within autonomous nations, will be asking for unity under a vast Asiatic despotism [sic]." (Quoted by A. Koshkish in his *The Sociopolitical Complex*, Oxford: Pergamon Press, 1979, p. 278).

6. For a defense of the questionable thesis that nationalism is a rebellion against the technocratic state see A.D.S. Smith, *Nationalism in the Twentieth Century* (New York: New York University Press, 1979); and J. Mascotto and P.-Y. Soucy, *Sociologie politique de la question nationale* (Montreal: Editions Albert St. Martin, 1979).

7. This diversity is well illustrated in Colin H. Williams (ed.), *National Separatism* (Vancouver: University of British Columbia Press, 1982).

8. On this, Atlan writes that "every people maintains an incestuous relationship with the land where it was born, which it wants to possess and to which it wants to return"; and thus "exile is like being torn away simultaneously from both mother and wife, like being without having been born" (*Entre le cristal et la fumée*, Paris: Seuil, 1979, p. 238, my translation).

9. See Yves Barel, *La reproduction sociale* (Paris: Anthrópos, 1973), pp. 448–449.

10. A recent and interesting attempt to introduce students to the study of politics from a similar vantage point can be found in A. Khoshkish, *The Sociopolitical Complex*.

11. These problems are actually the first two out of a list of five established by D.V.J. Bell (in his "Methodological Problems in the Study of Political Culture," paper presented to the 1974 annual meeting of the Canadian Political Science Association): i) "what are the 'elements' or 'components' of political cultures?", ii) "what are the relationships between 'micro' and 'macro' levels of analysis?", iii) "how is culture handed down or transmitted through time and across space?", iv) "problems of 'evidence' and 'inference'", and v) "what impact does culture have upon behaviour?" The third problem could advantageously be treated in terms of genetic analogy; this has been done by T.L. Thorson in particular (in Chapter 9 of his *Biopolitics*). But progress in this direction will depend on a further investigation of two crucial questions: how can we think of the information contained in the cultural code as a process of information and not merely as a set of instructions, and how can we theorize about societal evolution. The fourth problem is to a large extent a technical one that need not concern us here. The fifth and "most intractable" problem finds an answer in the notion of recursiveness which underpins this entire book.

12. See D.V.J. Bell and L. Tepperman, *The Roots of Disunity* (Toronto: McClelland & Stewart, 1980), p. 19.

13. See G. Vickers, "Values, Norms, and Policies," in *Policy Sciences* vol. 4 (1973), pp. 103–111 (the quotes are to be found on pp. 104–105). A more sophisticated analysis of these concepts along somewhat similar lines can be found in A. Khoshkish, *The Sociopolitical Complex*, Chapters 4 to 6.

14. See A. Giddens, *New Rules of Sociological Method* (New York: Basic Books, 1976), pp. 82–83, 86–87, 101, 208.

15. Predictability is not an absolute value; indeed knowledge about complex systems rarely allows one to make reliable predictions. But predictability is for behavioralists like Almond a cardinal value; thus to show that the study of political behavior does not enable us to predict future states of the political system is a relevant criticism.

16. See A. Touraine, *The Self-Production of Society* (Chicago: The University of Chicago Press, 1977, translated by J.W. Freiberg), Chapters 1, 4, and 5.

17. The seminal work in this respect is H. Eulau, *Micro-Macro Political Analysis* (Chicago: Aldine, 1969).

18. An exemplary study of bureaucratic policy-making is Colin Campbell, *Government Under Stress: Key Bureaucrats in Washington, London and Ottawa* (Toronto: University of Toronto Press, 1983).

19. The transition from collective to individual identities (or vice versa) is always delicate: is it possible to find a "model type" personality embodying a "national character"? This is a highly problematic line of inquiry and one which, for that very reason, has fallen into disrepute. But the enormous methodological obstacles raised by this kind of research should not be confused with a categorical refutation of its premise. See W.A. Rosenbaum, *Political Culture* (New York: Praeger, 1975), pp. 27–29.

20. *Supra* (Chapter 3).

21. See his *Aspects of Political Theory* (London: Collier Macmillan, 1976), pp. 131–143 and in passim.

22. Ibid., p. 143.

23. A. Etzioni, *The Active Society* (New York: Free Press, 1968), p. 473.

24. See Eric Nordlinger, *On the Autonomy of the Democratic State* (Cambridge, Mass.: Harvard University Press, 1981).

25. For a good summary and critique of this kind of policy studies see Joyce M. Munns, "The Environment, Politics and Policy Literature," *Western Political Quarterly*, Vol. XXVIII (December 1975), pp. 646–667.

26. Eric Nordlinger, *Autonomy*, p. 29.

27. As Marx wrote in the *Communist Manifesto*.

28. See R. Miliband, *The State in Capitalist Society* (London: Weidenfeld & Nicholson, 1969); N. Poulantzas, *Political Power and Social Class* (London: Verso, 1978); J. O'Connor, *The Fiscal Crisis of the State* (New York: St. Martin's Press, 1973).

29. See *Political Power*.

30. See *The Fiscal Crisis*.

31. For a contrary view, i.e., state sovereignty is an outdated and dispensable notion, see Paul W. Ward, *Sovereignty: A Study of a Contemporary Political Notion* (London: George Routledge & Son, 1928).

32. This example may be confusing insofar as it seems to invalidate my fundamental hypothesis regarding the organizing role of the state. But such is not the case because i) I consider the state only as the best instance of, and not as *the* organizing nexus within, a polity, and ii) for all analytical purposes, the party assumes traditional state functions so that if one reads "party-state" complex instead of state in which follows the argument would retain at least its descriptive validity (even if I still wish to submit that the "party-state" complex is a rather poor alternative to the sovereign state).

33. See D.D. Raphael, *Problems of Political Philosophy* (London: Macmillan, 1976, 2nd ed.), pp. 54–66.

34. T. Baumgartner, W. Buckley, T.R. Burns, and P. Schuster, "Meta-Power and the Structuring of Social Hierarchies," in T.R. Burns and W. Buckley (eds.), *Power and Control: Social Structures and their Transformation* (Beverly Hills: Sage, 1977), pp. 224–242.

35. See E.R. Service, *Origins of the State and Civilization: The Process of Cultural Evolution* (New York: W.W. Norton, 1975); P. Corning, *The Synergism Hypothesis: A Theory of Progressive Evolution* (New York: McGraw-Hill, 1983), pp. 368–375; G. Geiger, "The Concept of Evolution and Early State Formation," *Politics and the Life Sciences*, February 1985, pp. 163–170.

36. Jean-William Lapierre, *Vivre sans Etat?* (Paris: Seuil, 1977).

37. Ibid., pp. 75–76. Note that the Guyakis Indians whose political life is considered by P. Clastres as typical of stateless societies rank in fifth place in Lapierre's classification, insofar as they acknowledge the authority of a band chief.

38. Lapierre examines the relationships existing between the complexity of political organization and i) population size (ibid., Table I, pp. 156–157), ii) the development of productive forces (ibid., Table II, pp. 164–165), and iii) the degree of "ethno-cultural" homogeneity (ibid., Table III, pp. 170–171).

39. Space is lacking here to take into consideration Lapierre's detailed analysis of the unequal relations *among* African or American Indian societies (e.g., the creation of the Incan or Aztec empires).

40. See J. Habermas, "Toward a Reconstruction of Historical Materialism," in *Communication and the Evolution of Society* (Boston: Beacon Press, 1979), and also Chapter 5 *supra.*

41. *Op. cit.*, pp. 198–255. Incidentally, Lapierre blames Poulantzas for failing to see that the old feudal nobility had disappeared almost completely during the War of the Roses and that, consequently, it is wrong to pretend, as the latter does, that the revolution of 1688 marks the beginning of the transformation of the feudal nobility into a capitalist class.

42. See *The Sociopolitical Complex*, pp. 215–229.

43. Ibid., p. 226.

44. See *The Republic*, VIII 543A-550C on timocracy, VIII 550C-555B on oligarchy, VIII 555B-562A on democracy, VIII 526A-576B on tyranny.

45. *Legitimation Crisis* (Boston: Beacon Press, 1975, transl. T. McCarthy), p. 71.

46. See, for example, A. King, "Overload: Problems of Governing in the 1970's," *Political Studies*, vol. XXIII (June-Sept. 1975), pp. 284–296.

47. The following works are typical of this approach: A. Downs, *An Economic Theory of Democracy* (New York: Harper 1957); J. Buchanan and G. Tullock, *The*

Calculus of Consent (Ann Arbor: University of Michigan Press, 1962); M. Olson, *The Logic of Collective Action* (Cambridge, Mass.: Harvard University Press, 1965); A. Breton, *The Economic Theory of Representative Government* (Chicago: Aldine, 1974); and T.A. Borcherding (ed.), *Budgets and Bureaucrats: The Sources of Government Growth* (Durham: Duke University Press, 1977).

48. One of the best examples of this kind of reaction is the acceptance of "Proposition 13" by the citizens of California in 1978; its effect was to freeze property taxes in this state. It seems that taxpayers are more prone to protest when the economic climate is unfavorable, as it is in the 1980s, even if "objectively" the tax burden may be relatively light.

49. On this point, see F. Hayek's postscript ("Why I am not a Conservative") to his *The Constitution of Liberty* (Chicago: Gateway, 1960).

50. D. Donaldson, "Is there a Case for Minimal Government?", Paper No. P-84-5, B.C. Economic Policy Institute, The University of British Columbia (May 1984), p. 7.

51. F. Hayek, *The Constitution*, pp. 260–261.

52. See Peter Steinfels, *The Neoconservatives* (New York: Simon & Schuster, 1979); and Philip Abbott, *Furious Fancies: American Political Thought in the Post-Liberal Era* (Westport, Conn.: Greenwood Press, 1980).

53. See Steven Kelman, "Limited Government: An Incoherent Concept," *Journal of Policy Analysis and Management*, Vol. 3:1 (Fall, 1983), pp. 31–44.

54. A. Glucksmann claims that the political pathology of the 20th century results, in part, from Hegel's contrived attempt to reactualize what Glucksmann regards as Plato's betrayal of Socrates' thought; see *The Master Thinkers* (New York: Harper & Row, 1979), Part I.

55. See H. Ardendt, *The Human Condition* (Garden City, N.Y.: Doubleday, 1958).

56. For example, see Robert L. Kahn, "Americans Love Their Bureaucrats," *Psychology Today* (June 1975), pp. 66–71.

57. R.P. Hummel, *The Bureaucratic Experience* (New York: St. Martin's Press, 1977), pp. 96–97.

8

Political Theory and the Evolution of Ideas

What societal actors know about the natural and societal orders influences their relation with one another and, over time, contributes to shape civilizations. The *critical* examination of this process is one of the tasks of political theory. In turn, political theory informs historical development. Indeed, in order to complete the exploration of the three dimensions of political life, I still have to account for, and evaluate, the ways in which political actors construct their public realm according to their self-consciously articulated visions of the ideal polity. What is needed in this respect is a meta-theory of the complex interrelations existing between ideologies, science, including the social sciences, and normative political theory. Such a meta-theory is, and will continue to remain an elusive goal but one that we can hardly afford not to pursue. Two avenues of research are proposed here. One is a reflection on the autonomous and self-reforming character, not only of the political actors themselves, but also of all biological beings; the other focuses on the "ecological" context of autopoiesis[1]—"eco-logical" connoting here constraining pressures external to the members of a polity as well as an active intervention in the management of the collective household, including much of the biosphere. The ontological and epistemolotical implications of the self-referential logic inherent in our human existence are discussed in the first section, and further developed in the following section where a strategy for articulating meaningful metaphors expressing the autonomy and subjectivity of political actors is outlined. Finally, the ecological aspects of the reproduction of the polity are discussed in the last section.

The Rediscovery of Subjectivity

It is hardly necessary to dwell on the multiple relations between politics and ideology. But it is only recently that political theory has posed the problem of the complex interdependence of science, ideology and politics, which is so characteristic of our advanced industrial civilization. (Of course, the sociology of knowledge has prepared the ground for an investigation of this kind, but we need to focus on the specifically political dimension of knowledge.) The isolated ivory tower of science is actually a pillar of the dominant (technocratic) ideology. It is being maintained, at an ever increasing cost, by armies of bureaucrats both without and within its walls.[2] Technological control over nature is the mirror image of bureaucratic control over society. On the one hand, the post-Heisenbergian natural sciences have revolutionized our understanding of the physical universe which we no longer view as being totally independent of the scientific observer.[3] On the other hand, scientists fight hard to preserve the fiction of objectivity and neutrality when they have to account for the social relevance of their work,[4] thus reverting to a positivistic paradigm that many of them already have rejected in their research. Scientists, regardless of their own epistemological convictions, have an interest in the defense of scientism. A Baconian interpretation of the scientific enterprise, in conjunction with the actual results of "scientific progress," can legitimize science much more effectively than contemporary scientific epistemologies could ever do. To continue to promise prosperity and the security of living in a domesticated nature is the best strategy for insuring that science will remain at the apex of the hierarchy of values. A self-critical, humanized, perhaps even "Buddhist," science[5]—one which is already practised, if in a somewhat unofficial way, within the scientific community—would find it much more difficult to establish its credibility in the market place.

This scientific "double speak" ought to be a dominant question for political theory. As Edmund Husserl explained:

> Along with growing more and more perfect cognitive power over the universe, man also gains an ever more perfect mastery over his practical surrounding world, one which expands in an unending progression. This also involves a mastery over mankind as belonging to the real surrounding world, i.e., mastery over himself and his fellow man, an ever greater power over his fate.[6]

Unfortunately, both the Marxist and liberal schools of political theory rest themselves on scientistic assumptions regarding the availability of natural resources and man's trans-natural status (i.e., both define nature as pure externality).

We urgently need a meta-science of politics which would take the form of a critique of the ideology of science, of scientism and of political theory. Its theoretical domain is sketched out in Figure 8.1. Such a critique could usher in the age of what Raghavan Iyer calls "parapolitics," by which he means a practice that "emphasizes the architectonics of the progressive awakenings and movements of humanity within the evershifting relationships

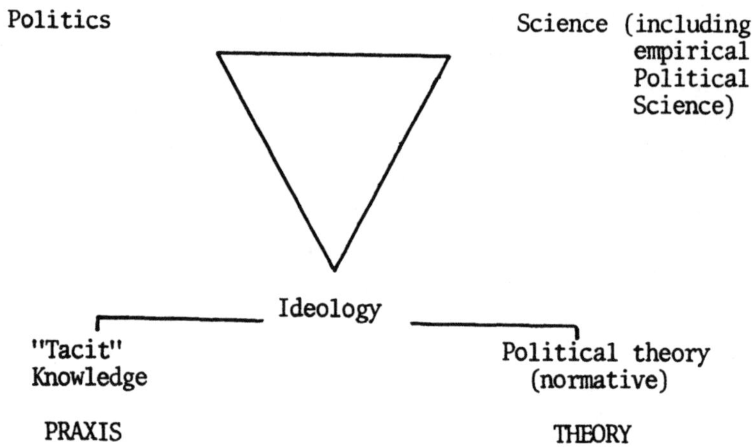

Figure 8.1

Theoretical Domain of a (Meta-)Science of Politics

of the One and the Many, the whole and the parts."[7] A para-political reflection on the post-modern trilogy of science, technology and bureaucracy would illuminate the tension that could bring about an entirely new political consciousness. As science confronts complexity more directly, it reveals a typically contemporary tension between rational knowledge and the chaotic unknowable. Technology is fraught with an inner tension between doing and being, that is, between the urge to transform the condition of our life and the need to experience and contemplate these conditions before they change beyond recognition (e.g., the politics of environmental protection). And the bureaucratic world is torn between the collective and individual dimension of our political selves, as we try to reconcile the requirement of efficiency with our understanding of the practical implications of human rights (e.g., the controversy over affirmative action).

By definition, any meta-level remains beyond our reach, for, as Pogo said, "I've seen the enemy and it is us!" Parapolitics is an end that will always evade us. However, in striving toward such a goal, we may develop a new vision and reach a new understanding of ourselves. In what follows I examine ways of actualizing this tension, beginning with a reflection about the ontological basis and epistemological strategy appropriate for a post-positivist science.

The Epistemological and Ontological Loops

Man's separation from nature has been accomplished in several stages, beginning with what Habermas calls the period of "demarcation" in neolithic times.[8] But the formal divorce between man-the-historical-subject and nature-as-exteriority was pronounced in the 17th and 18th century, most notably by Descartes,[9] Hobbes,[10] and Rousseau.[11] We have come now to realize the full extent of the consequences that follow from this divorce. Without it, the natural sciences would probably not have escaped the strictures of scholasticism. But the negative consequences appear, with hindsight, to have been equally momentous. One of these consequences is that we find it very difficult to express the likeness of politics with life because the naturalness of the polity has been eclipsed by its historical uniqueness; we are left with only mechanist models of nature (including mechanicist biological models). When applying these mechanicist models to political inquiry, the temptation is great to bracket out the historical subjects altogether (e.g., behavioralism) thus reducing man to an aggregate of socio-economic data and politics to the operation of the "political system." In reaction against such a tendency, some prefer to interpret all forms of knowledge (including physics) as being historically determined ideologies, which also amounts to a reductionist position.

In a symmetrical fashion, the natural sciences also borrow concepts from the anthropo-sociological and political domains. There appears to be, or at least it is tempting to see, a correspondence between the order of Newton's cosmos and the rigidity of the social hierarchies of his age, on the one hand, and between the chaotic universe of quantum physics or Prigoginian thermodynamics and the chaotic transition to a post-industrial age through which we are going at present, on the other hand.[12] Moreover, as E. Morin observes:

> The terms communication, information, code, program, message, finality, have emigrated from anthropo-social experience into the cybernetics of artificial machines and, from there, into [the domain of] biological organization and come back under their new cybernated form to invade [the field of] anthropo-social organization.[13]

It is not the circulation of ideas from one discipline to another but its "clandestine" character that Morin objects to. Let us accept the inevitability of these conceptual transactions and work towards their organization in a reflective and critical manner. The first step in this direction, as E. Morin aptly suggests, is to move away from the reductionist tendency to conceptualize knowledge transfers in linear terms. Rather, it is preferable to visualize a recursive flow of ideas along an "epistemological loop" (see Figure 8.2).[14] Morin expresses the hope that this "recursive network, where the socialization '*physis*' and the physicalization of society would become co-producer of each other, . . . ought to constitute the principle of a new theoretical vision." And he adds that such a vision would ultimately lead to a "double theoretical

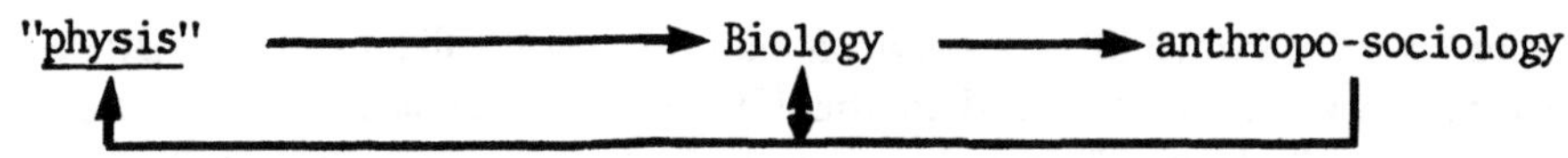

Figure 8.2: The epistemological loop

(Source: E. Morin, La Méthode, vol. 1, p. 17)

grounding in 'nature' and in 'culture,' in the object and in the subject."[15] I wish to elaborate upon the latter connection for it represents, at least in my view, one of the best points of entry into the epistemological loop or, to use another analogy, one of the most promising strategies for exposing the "clandestine" conceptual transactions and thus for applying normative discourse to bear upon their significance (or lack thereof).

The Co-Production of the Subject and the Object

There are many ways of approaching the question of how to ground the epistemological loop into the co-production of the object and subject. This is a metaphysical question *par excellence*, but I will allude only to the methodological aspects of such an inquiry before discussing its political implications at some greater length.

The question of the relation of the observer to his observations has been discussed by many authors. At the risk of oversimplifying things, there is an on-going debate between those who investigate the *psychological* context of the observer and those who are primarily concerned with "the context of *validation*." The former ignore the physical and biological substrata of the human mind whose specificity can only be appreciated by reference to the (objective) knowledge we have acquired about the brain; the latter dismiss the fact that validation is a linguistic act and that language, more than any other human creation, manifests the irreducible subjectivity of the human mind. These questions become even more complex when the observer is a more or less active participant in the reality under observation. I argue that political actors produce their public realm and are, in turn, reproduced and nurtured in ways that Morin's epistemological loop reveals more completely than alternative approaches. (Immediately hereafter, I critically examine two of these alternative approaches while I analyze the positive import of Morin's ideas in the next subsection.)

It is not necessary to link politics to natural phenomena in order to comprehend the co-production of political subjects. For example, Anthony Giddens has proposed a lucid analysis of the partial (and problematic) overlapping of the micro- and macro-political processes resulting from the fact that "rules and resources are drawn upon by actors in the production of interaction, but are thereby also reconstituted through such interaction."[16] His theoretical program (i.e., what he calls the theory of societal "structuration") converges in some important respects with the theory of self-organizing systems—a convergence which is amply demonstrated by the numerous references to his work found throughout this book. However, Giddens remains largely silent about the naturalness of societal processes—except when he criticizes the naturalistic assumptions of functionalism and General Systems Theory.[17] This is all the more regrettable since he has perceived the need for something like Morin's epistemological loop:

> The main implication of the ideas I have set out in this paper is that, in the current phase of social theory, we are involved in rotating two axes simultaneously: that of our understanding of the character of human social activity, and that of the logical form of natural science. These are not entirely separate endeavors, but feed from a pool of common problems. For just as it has become apparent that hermeneutic questions are integral to the philosophical understanding of natural science, so the limitations of conceptions of the social sciences which exclude causal analysis have become equally evident. We cannot treat natural and social sciences as two independently constituted forms of intellectual endeavor, whose characteristics can be separately determined, and which then subsequently can be brought together and compared. Philosophers and practitioners of sociology must remain attentive to the progress of the natural sciences; but any philosophy of natural science in turn presupposes a definite stance towards problems of social theory.[18]

Recently, some authors have embarked on a research program aimed at exploring the interconnections that exist between biology and political science. They started off from an interesting idea, namely that the scientific study of politics must take into account the genetic and physiological as well as the socio-economic conditions of political life. There is a risk, however, that somewhere along this road we may lose sight of the contribution that political actors (subjects) purposefully make to the construction of the public realm (i.e., the sum total of political objects).

Biopolitics, as this new direction of research is sometimes called, is not a unified approach.[19] According to Albert Somit,[20] there are three types of bio-political problems: i) the application of the concepts and methods of ethology to the study of political behavior,[21] ii) the analysis of the physiological aspects of political behavior,[22] and iii) the investigation of the public policy applications of biology (e.g., genetic technology).[23] I see no methodological problem of great significance in the case of the latter type of research; the other two raise more serious questions that I address below. Somit's typology though is not complete. We should add also the expanding research field consisting of a variety of evolutionary frameworks integrating biological and socio-cultural evolution.

The analysis of the physiological aspects of political behavior is a challenging and promising research field, but one which is fraught with formidable methodological pitfalls. To ignore the somatic basis of political behavior would be like imitating the proverbial ostrich and burying our head in the familiar sand of psychological and societal factors pretending that human beings have escaped the confines of the biosphere. Even if one wished to do so, it is bound to become an impossible situation in view of what has been learned so far on neurophysiology and the structure of the brain, for example. Glendon Shubert has surveyed an already impressive literature on the relationship between the structure of the human brain and political perception, between memory, learning and political thinking, and on brain development and political equality.[24] A recurrent theme in several of these research projects is the possible influence of the lateralization of the brain (i.e., the left hemisphere being more oriented toward analytic operations, and the right hemisphere more holistic) on a number of politically relevant factors, such as conflict resolution, or the degree and extent of hierarchical structuration in social organizations. A most intriguing, and ideologically charged question in this respect concerns the correlation between sex and certain neurobiological patterns conducive to distinct patterns of political behavior (e.g., orientation toward defense, quality of the environment, etc.), since there exist significant developmental differences between most male and female individuals as far as brain lateralization is concerned.

However, political scientists are well advised to engage in this kind of research with the utmost caution. It is salutary to be reminded by a practicing neuropsychologist that "it is premature to say anything profound concerning any neuropsychological research finding and how it might relate or be incorporated in political science."[25] While this may be an overstatement, it remains that i) political scientists, by definition can only hope to be gifted amateurs dabbling in the interpretation of laboratory experiments that they are most likely to be rather unfamiliar with; and ii) the methodological problems associated with the identification and operationalization of the intermediary variables relating physiological factors to political ones are so immense that progress in this direction is bound to be frustratingly slow.

Moreover, the same kind of criticism that was directed at pre-biopolitical behavioralism can be made with respect to biopolitical behavioralism, namely that there is more to political life than the predisposition of individuals likely to react in similar ways to psychosocial or biological constraints. For example, it is certainly not trivial to have discovered that power-seeking individuals, as defined by their score on a questionnaire developed by researchers interested in the relationship between aggressive/competitive behavior and coronary disease, have higher levels of whole blood serotonin (or 5-hydroxytryptamine)—as do dominant vervet monkeys.[26] But to what extent is this *politically* relevant? My answer is: only to a limited extent. For the politically significant question is not: Are there individuals who are more predisposed to dominate other individuals in a social context? It is rather, given the common sensical realization that this is certainly the case, even if the somatic basis of this truism was not known until recently: What means do power-seeking individuals use, and to what ends? It is precisely because he answered this question in a compelling manner that Machiavelli is still read today.

Ethology and evolutionary models are more obviously relevant to political inquiry. As I have discussed already at some length many questions related to these cross-disciplinary linkages, I do not have to elaborate on this point. Nevertheless, this is an opportunity to restate succinctly my position. If we accept that there are no facts outside of the theoretical frameworks within which we construct images of nature and society, it is clear that the life sciences in general, and ethology and evolutionary biology in particular are invaluable building blocks for political scientists whose vision of political phenomena is not limited to Skinnerian behaviorism or dogmatic cultural determinism. (Specifically on the question of the import of ethology, I concur with R.D. Masters' observation that the organism-stimulus-response (O–S–R) model inherent in ethology "appears to be more general and more fruitful than the stimulus-response (S–R) framework of American behaviorist psychology."[27] However, my contention is that, if we are to move beyond these paradigmatic explorations and into the field of empirical research—and I feel no urge to hasten the pace in this direction—we must have a very clear notion of what we are trying to do. I suggest that we ought to use our knowledge of animal societies and sociobiological dynamics in order to delineate and emphasize the uniqueness of human life, and the specificity of political phenomena. Valuable as they may be, studies of political behavior which emphasize the relevance of ethological findings without also setting limits to their use in interpreting political phenomena are unsatisfactory, if not mystifying. For example, territoriality may not be—in fact, is not—a specifically human trait, and T. Wiegele's advice that more work "needs to be done by political scientists to explore comprehensively the biopolitical roots of human territoriality"[28] is probably well taken. But territoriality is only one aspect of nationhood, and today perhaps not as significant as it was in the past. In other words, ethological data are useful if they serve to further underline the complexity of political life. However, they can be misleading if they are submitted as a ready-made explanation or simplification of complex relations. By contrast, R.D. Masters' analysis of the import of inclusive fitness theory for political science is particularly insightful in that it underlines the uniquely human propensity to co-operate with genetically unrelated individuals for the purpose of achieving mutually beneficial results; Masters also points to the complexity of the calculations that lead to this kind of extraordinary behavior.[29] What then is the task of political theory? I submit that it consists in revealing both the objective condition under which political institutions are created in order to organize co-operation, and the meaning of the political actors' involvement in these institutions.

So far, I have equated the notion of subject with individual political actors; this seems indeed to be the dominant interpretation. However, there is another aspect of the co-production of political objects/subjects that remains generally overlooked. We have seen in the last chapter that a polity can also be regarded, at least metaphorically, as a personality—a complex, self-contradictory and fuzzy one but a personality nevertheless. I suggest that in order to comprehend the co-production of the macro political objects/subjects we also need to adopt an encyclopedic epistemological perspective.

The state is an objective reality that manifests itself as both the cause and the effect of the actualization of the potentialities inherent in the national group(s) and other collective subjects which produce it. The state negates its objective nature by constantly trying to become what it cannot be: an omnipotent and omniscient societal being. The modern state comes rather close to being omnipotent in that it intervenes in almost all areas of social life—and in so doing assumes responsibilities (e.g., environmental protection) that make it almost as much a part of the biosphere as it is a political entity. It is also in a sense omniscient because science has become institutionalized around the state, so to speak, and because the modern state is archetypically a technocratic state.

To sum up, a fundamental problem for political theory is that of conceptualizing highly complex political phenomena partaking of several orders of reality in such a way that politics is not reduced to some infra-human order of process nor, inversely, in such a way that the experiential basis of the natural sciences remains unexposed. I have attempted to establish that, in order to tackle this problem, empirical theory must be grounded in an ontological position predicated upon the recursive interdependence of the objective and subjective halves of the physical world, of the biosphere, and of the public realm we inhabit as physical, living and political beings. Next, I examine how we can express such a concern in the very way we organize political discourse.

Life and Political Life: The Search for True Metaphors

The Indispensability of Metaphors in Political Discourse

The problematic linkage that societal actors produce and reproduce between modes of action and political knowledge is typified by the difficulties inherent in the derivation of political metaphors from the natural world. First, it is necessary to explain why political discourse resorts to metaphors. My answer to this question parallels Northrop Frye's attempt to show that Biblical interpretation requires a judicious usage of three distinct levels of language, i.e., the metaphorical, "metonymic" and "demotic" languages.[30] Each one of these levels has its limitation but they are *all* constitutive of the whole. In other words, metaphors are not the only, and certainly not the best linguistic account of political phenomena. However, no satisfactory account of politics can be formulated which does not include metaphors. The problem is to find a way of articulating those images that provide us with clues for understanding the complexity of political phenomena, and, of course, to avoid reification, that is, the uncritical use of what M. Landau calls "rigidified" metaphors.[31]

A metaphor consists in the transposition of a concept/word from one specific context into another distinct but comparable one. Thus a metaphor is a linguistic operation of the A-is-B type[32] (e.g., "power is an aphrodisiac"). From a chronological standpoint, metaphors are the oldest form of expression. In Socrates' times, the second phase of language (i.e., "metonymic" language) began in which words are "'put for' something that by definition transcends adequate verbal expression"[33] (e.g., the line that we draw is "put for" the ideal line in Euclidean geometry). This transition to more complex analogies allows the expression of logical connections between concepts abstracted from their descriptive context. Finally, with the Renaissance, descriptive or "demotic" language became culturally ascendant; it is "a mode of thought and speech in which the word is 'put for' the object it describes."[34] Implicit here is the separation of object and subject and, in particular, the assumption that objects can be perceived and conceived as they are, independently of the cognitive and linguistic structures characteristic of the subject who formulates an empirical description of external reality (e.g., the earth is a sphere).[35] Modern political theory avoids metaphorical language but, just as theological exegesis is not usually capable of giving a proper account of the poetic and mythical dimension of the Bible, the objectifying language of political *science* cannot express the complexity of political phenomena, for metaphorical and mythical expressions abound in ordinary political discourse. For example, we talk about "political life" without reflecting upon the meaning of such a phrase. It is my contention that a reflection of this kind would be worthwhile. It is not "just a metaphor," it is a very suggestive and powerful one. "Political life" can be understood as an element of what Ludwig Wittgenstein called "language games," that is to say games whose rules permit us to see the truth of a statement by reference to criteria derived from social practice rather than from a formal and disembodied correspondence with an objective reality. The importance of practice in evaluating the significance of metaphors is also stressed by E.F. Miller.[36]

Although the most active promoters of the behavioralist paradigm did themselves employ more or less explicit metaphors in the construction of their theoretical frameworks,[37] they have criticized the grand theorists for the immoderate use of "unverifiable" metaphors. According to their "verificationist" viewpoint, to use E.F. Miller's term, analogical reasoning constitutes an acceptable methodology only if it creates models whose structural correspondence with a non-political phenomenon can be tested empirically.[38] Miller's well-taken criticism of the verificationist position is that no empirical test would ever satisfy such a requirement. The reason being that "political behavior" is not directly observable but must be linguistically (and thus, to some extent, metaphorically) inferred from more directly observable types of behaviors.

The "constitutivist" view, on the other hand, denies the existence of a reality that would not be dependent on the act of describing it through language. And because there is no language in which metaphors are not used (except for pure logic), they are an unavoidable linguistic practice in political discourse.[39] While I think that such a position is defensible if it is not dogmatically enforced, I concur to some extent with Miller's criticism of it on the grounds that, even if some arbitrary operation of demarcation is inherent in scientific observation, provisions must be made for the existence of social relations and ecological constraints occurring independently of the act of describing them. Relativism, I would add, may lead to an evolutionary cul-de-sac. (Yet I am prepared to admit that "external" reality is nonsense, because it is always internal to the actors who produce that reality and make sense of it for themselves, at least.)

Thus we are faced with a dilemma: How can we treat metaphors as essential constitutive elements of political knowledge without reducing the latter to the former? Miller's answer is to adopt what he calls a "manifestationist" perspective. I consider his solution to be basically sound albeit somewhat limited in scope. He argues that metaphors help us to realize the significance of general principles of likeness between the domains of immediate experience and politics. However, these metaphors still have to be tested, not empirically, but through rational criticism:

> The closer we approach an integral or comprehensive view [of the domain of politics], the more fully we are able to appreciate both the truth that each metaphor makes manifest and the partiality of that truth. Political experience whose significance the metaphors have brought to light can now be used to test metaphorical perspective and ascertain their limits.[40]

My only criticism is that I do not consider it to be sufficient to limit our vision of the "whole" to politics alone, even if such a vision may already seem elusive enough. Metaphors are also essential to the construction of natural scientific theories. The whole that the interplay of physical, biological, and socio-political metaphors should render manifest is the evolutionary pathway of the "political animal" considered in his animal as well as in his political nature. I intend now to explicate why and how an encyclopedic perspective that reveals the never resolved tensions between autonomy and ecological dependencies is required.

All Machines Are Not Machines . . .

In E. Morin's recent writings we find an abundant mine of analogies that meet the standards defined immediately above. Travelling back and forth along the epistemological loop, Morin has perfected the concept of autopoietic machine whose relevance I have illustrated already in my discussion of the epistemological foundation of revolutionary biology. Artitifical machines provide us with a useful starting point for the exploration of the physical universe. Indeed, the invaluable contribution of classical mechanics—as paradigm which cannot be divorced from the mechanicist world-view to which it gave rise and, in turn has come to depend upon—revealed, and formally expressed the energy transactions that take place among physical systems. (And, in a sense, the polity is one of them.) But it has become quite clear that the various machines producing and reproducing our universe (e.g., the stars at the one end, the atoms at the other) possess (re-)generative properties lacking in artificial machines. (Note, however, that we may learn, in the very near future, to produce on a large scale semi-artificial autopoietic machines by means of genetic manipulation of the genetic code of bacteria!)

The methodological moral of this cosmological tale is that artificial machines, far from being perfect archetypes of the universal laws of motion, are but poor models of the natural autopoietic "arkhe-machines" (Morin). *A fortiori*, the mechanicist perspective reveals little about political phenomena in general, and practically nothing about the self-organization of the public realm. However, it helps us to understand the transformative, and the productive (but short of the re-productive) aspects of political life (e.g., Easton's input-output model, or Hobbes' Leviathan which channels human "e-motions"[41] for the production of a stable and integrated body politic). What natural machines teach us is that productive activities can be brought to bear upon themselves. This is not exactly a discovery for the social sciences, but it relativizes the specificity of societal reproduction and poses anew the problem of the nature of the societal self.

The transition from the clockwork to the self-renewing *physis* also means that technology, and human labor in general, need no longer be *the* basis of societal reproduction (although it obviously remains *a* basis). Rather, it is the totality of human life that is involved in the recursive closure of productive activities upon themselves. Now life, from cells to man, manifests its self-organizing properties much more fully, and indeed much more openly than any physical system, as the concept of autopoiesis clearly illustrates. At the same time, the paradigm of the autopoietic machine does not account for the totality of the phenomenology of life, precisely because this paradigm fits the living organism so perfectly. I mean that we discern the emergence in living organisms of properties that, even though they are not sufficient or even necessary for understanding the behavior of living systems, account for the discrepancies that we intuit between our implicit, not to say naive, understanding of life and the autopoietic model. Briefly stated, the identity of living organisms is founded not only on a fully developed dis-junctive operation of *self*-recognition as it is exemplified in immunological reactions, but also on an emerging egocentric reference that marks, in Morin's term, the transition "from self to me"; moreover, this "me" is experienced by the living organism as an "I."

This demonstration could be carried out rather easily at the level of multicellular organisms; after all, chimpanzees and gorillas have been taught the ASL language and thus have evidenced that they can manipulate the linguistic categories "me" and "I."[42] The logic of the demonstration would become admittedly more metaphorical but it would still be rather uninstructive, although not irrelevant, if it concerned some lower species. But, in a provocative and original manner, Morin tracks down the origin of these categories to the level of unicellular organisms.[43] In so doing, he is led to formulate systemic principles that, in turn, can be readily applied to the analysis of the development of the individuality inherent in human organizations.

The fact that the genetic code is a program and not a mechanistic design or blueprint,[44] i.e., that it is a set of instructions which must be "computed" rather than a direct causal determination, allows the cell to proclaim, as Morin facetiously explains, "*computo ergo sum*."[45] The actualization of a program requires the mediation of a computer, in the generic sense of a system that com-putes, i.e., a system that compares and evaluates data and selects an alternative by reference to the norms specified in the program.[46] The cell is a computer that never stops to compute until its death, chemically processing information about its proteinic make-up and external stimuli, and selecting responses that are appropriate, not because they are predetermined but because they are compatible with a genetic instruction.[47] Consequently, Morin explains:

> . . . the self-referential network from self to self brings back the computed in the computer; the computed being also the computer, the computer-computed comes back in the computation of the computer. We have here a self-cognitive network where the computer is capable not only of computing

> its own parts through the whole of itself, and computing itself as a whole through its parts, but also of objectivizing itself as computed ("Self", "Me") and resubjectivizing itself as computer ("I"). In other words, such a process supposes the constituting, distinction and identification of an "I" (egocentric computer), of a "Self" and of a "Me" (objective computed entities that are identified to one another and both to "I").[48]
>
> Thus we see that the self-referential logic of the self-cognitive network requires that the subject recognizes itself under the three mutually necessary instances that we express by the terms "I", "Me" and "Self". The "I" refers to the occupation of the egocentric locus by the computation—the "*computo*"—carried out by a being asserting itself as an individual-subject. The "Self" is like the physical corporeality of the "Me-I". The "Me" partakes both of this physical corporeality and of the invariance of the "I".[49]

Now the paradox is that, while the cell is a cognitive system that knows how to compute itself, as it were, it does not know what it knows, it is not cognizant of itself. This reflective knowledge can only be achieved by polycellular organisms endowed with a central nervous system. And the ability to make choices, to work out strategies, reacts in turn on the cellular "computation."[50] Man, of course, has perfected this ability to an uncomparable extent (the human brain-mind loop constituting yet another recursive level).[51]

Thus the organism is not a coherent structure wherein the parts neatly fit in the whole. It is a complex, at times even contradictory, relationship between two separate but interdependent orders of individuality. (Indeed ethology has clearly established that animals are more than the sum of their genes.) This is something that the simplistic organismic analogy (upon which functionalism is predicated) cannot account for.

Morin's suggestive metaphors contribute to our understanding of the life of the political animal in at least two ways. First, we are compelled to sharpen our analysis of the subjectivity of the political actor (subject). Reductionist theorists sometimes pay lip service to his autonomy but rule out its effects as amounting only to marginal variations with respect to a clearly defined behavioral pattern. In other words, they are content with the notion that political actors organize their lives as egocentric atoms of action, just as unicellular organisms build up their sense of "Me/I." But they feign to ignore that the political actors *know* what they are actually doing. This is not acceptable for humans are no mere bacteria! Not only do human beings know how to organize their own self-centered lives, but they also understand something about the true import of their actions. The metaphorical description of political phenomena as political life forces us to pay attention to the specifity of political action, since individuality is common to all forms of life. This is the point from where an exploration of the subjectivity of the political actor must begin. Empirical political theory, from this standpoint, is not about actual behavior but about the context of behavior, as perceived by, for example, decision-makers who have been given access to voters surveys, economic forecasts, social impact studies, etc. But the problem is even much more complex, for as A. Giddens reminds us, "every social actor knows a great deal about the conditions of reproduction of the society of which he or she is a member."[52] Therefore, to speak of either empirical or normative political theory is inappropriate: political theory is *critical* theory. Its purpose is to confront the actors with not only the unintended, but also the unavowed consequence of their action, to challenge their motivation, and to assess the significance of the norms that are applied to the evaluation of the available options. But critical theory is not a substitution of the theorist's insights for the flawed and crude practical knowledge of the societal actors, since critical theory is dependent upon the practical knowledge of both the "doers" and the "done-by" to produce its desired effects. Again, the political actors' subjectivity, their sense of self, is not coterminous with their (biological) individuality. Political life is the complex life of the polity as a whole. To illustrate this point, let us analyze the on-going debate between the advocates of *dirigisme* and the defenders of laissez-faire. The latter are correct in pointing out that there exists no scientific basis upon which comprehensive planning schemes can safely rest, since the predictive power of social scientific models is, at best, limited. However, the alternative notion of a "spontaneous" social order, proposed by F. Hayek and others,[53] is also deficient insofar as it is meant to provide a symmetrical foundation for an objectified social practice—what L. Allison calls "marketism." The spontaneity of economic behaviors is a very problematic concept. Certainly, societal actors do not conciously reconstruct the economic structures within which they live before engaging in various economic activities, and do in fact react tacitly to the incentives and constraints which they are aware of. However, this tacit, practical knowledge includes not only criteria reflecting their own interests, but also cultural norms. Even more importantly, they also reflect either the dominant or the challenging social theories and ideologies to which social actors are confronted and in relation to which they define their own values in a not-so-spontaneous way. Marketism is precisely a system of ideas of this type. While egocentric amoebae spontaneously identify their own locus in their ecosystem, political actors enter into much more complex relations with their alter egos, relations which are mediated by political theory and further influenced by the critical reevaluation of accepted theoretical claims.

My second and related point is that the metaphorical emergence of a societal being modeled on the development of an autonomous organism consisting of computing cells can be used for furthering our comprehension of the tensions existing between individual and collective values. The polity may be a meta-individual but it ought not to be conceptualized as a mega-individual (in the same way that we now know that cells are not mere cogs in the phenotypical wheel!). For Morin, animal societies have a weak "ontological density";[54] they are self-organizing but not completely individualized. Let us recall that the conditions of the emergence of the biological subject are the existence of a generic (genetic) *logos* (code) which is interpreted (computed) by an egocentric being (computer). Only human societies meet these conditions insofar as they have developed cultural codes that are actuated by human minds/brains (and in turn react on the latter).[55] The nation-state is a meta-structure emerging as a consequence of the communications that individual actors enter into as they individually realize their cultural heritage (a heritage which is, moreover, unequally distributed). This public dimension offers us the possibility of self-transcendence and thus of giving meaning to our egocentric subjectivities. It also contains the inherent possibility of domination and exploitation. In brief, the naturalness of society is to be understood as a meshing of interacting and yet distinct subjectivities.

The subjective dimension inherent in life had never completely escaped laymen but its rediscovery by at least some biologists (e.g., F. Varela) and its clandestine reappearance in the theoretical framework of modern biology (e.g., ethology) are breakthroughs that warrant the attention I have so far devoted to them (and to Morin's exegesis of their ultimate significance). However, all this is not to suggest that life is pure subjectivity. The logic of life is doubly dialectical: tensions develop between, on the one hand, the invariance of the genetic code and its actuation in epigenesis and, on the other hand, biological subjects and their environment. Life obeys both *auto-* and *eco*-logical principles, to use Morin's phraseology.

The prodigious complexity of the self-organization of eco-systems provides us with a wealth of analogies of a very different kind. This time I turn my attention to the constraints relativizing the autonomy of living systems and, in particular, man's relation to his natural environment.

The Political Economy of the Autopoietic Polity

The Political Animal's Double Nature

In order to understand something about human nature we have to follow a winding path through the physical, biological, and anthropological realms: that much has been established so far. But are we actually fit to undertake this cosmic journey? Of course, the very fact that we exist as living matter allows us to relate intuitively to matter and life. However, the relevant question is whether or not we have the practical knowledge required for traveling safely along Morin's epistemological loop. For example, it is a matter of great concern to some people that we do not seem to have learned well enough how to interact with nature without destroying it or without destroying ourselves. The paradox of modernity is precisely that we have gained our much vaunted knowledge of nature in a manner that may lead ultimately to the ruin of our technological civilization, if not of nature itself.

Is this a question for political theory? There does not seem to be any direct relationship between the function and the reproduction of the polity, on the one hand, and man's "control" of his environment. Or if there is, it is often presented in antithetic terms: global concerns (ought to) supersede particularistic and cultural preferences, whereas politics stands in the way of the disinterested scientists' efforts to save man from himself, or so we are told by the Club of Rome and similar organizations. However, I intend to argue that the problem of man's relations with nature cuts across, and thereby reveals the very essence and the actual practice of political communications in post-modern societies. This inquiry makes it possible to reflect upon the ends of politics in a theoretically and practically relevant manner.

Some may retort that this is nothing but a belated rediscovery of what Aristotle had intuited long ago: man is a political animal (the emphasis being now placed on the *second* term). But the meaning of Aristotle's phrase remains somewhat ambiguous; it is probable that he himself did not realize the extent to which his explanations were inconsistent. On the other hand, it is the very ambiguity of his writing that makes it interesting. Let us, therefore, briefly examine Aristotle's position, not so much in order to "solve" those ambiguities but in order to use them creatively for the generation of new hypotheses about the reproduction of the polity.

R.G. Mulgan has shown that Aristotle used the phrase "*politikon zoon*" in three different though related senses.[56] First, man shares with other animals the need for the company of other members of his species. In other words, man is a social (and sociable) animal. (In this respect we have much to learn from ethology.) Second, man needs to participate in collective activities of a specifically human type, inclusive of political activities but not limited to these. Finally, man is also a political animal because he has a need for the "polis" exclusively of other social institutions, politics being the condition of his self-actualization. At first sight, it may seem as if the latter sense is incompatible with the first two others; if so, Aristotle left little doubt about the paramount importance he attached to man's bond to the polis. Yet it may be instructive to reflect upon the various ways in which these various definitions can be related to each other—not in order to achieve some kind of bland synthesis but rather so as to better understand their paradoxical unity.

To say that man is a member of the polis does not—or should not—reduce him to his political self. As Aristotle conceded, man belongs to a variety of social institutions. His insistence on man's need for the polis exclusively of other social institutions can best be understood from a diachronic perspective. Even though societal actors may, at any time, exchange goods and services, or create and destroy cultural values, without engaging in political communications, the perpetuation and even the growth of these exchanges inevitably raise questions that can be settled only in the political arena (from the Aristotelian standpoint, this expresses the tension between actuality and potentiality). Now, many intra-societal exchanges also imply transactions with the biosphere and/or inanimate nature. Whereas political theory proper allows us to make sense of what is said in the political arena (and to take part in the debate), political economy, as it was understood by its founders, allows us to understand the relationship of these debates to the perpetuation of economic interactions. However, we seem to have lost the means whereby we could articulate in theoretical discourse the problems that we face daily but privately, and therefore somewhat blindly, in our relations with the natural environment. The reason why I write "lost" is that the short lived physiocratic school was precisely concerned with this question. Fortunately, René Passet recently undertook to examine the conditions under which a revival of this tradition could be achieved.

Toward a Neo-Physiocratic Political Economy?

Morin's epistemological loop itself subsumes a variety of isomorphic linkages among cognitive domains. Within the "anthropo-social" pole we can identify an economics—political science loop, i.e., political economy. Moreover, the linkage between biology and the anthropo-social pole includes a connection between ecology and economics which has become the focus of more attention. Thus I will be concerned hereafter with the following loop.

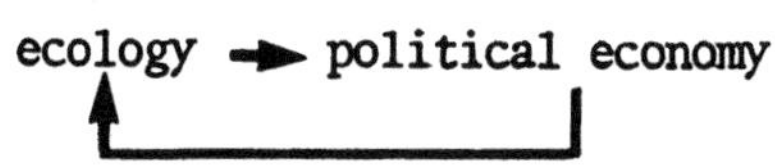

According to R. Passet, this paradigmatic vision was given an actual theoretical formulation once in physiocracy, i.e., in the works of Quesnay, Turgot, and Le Mercier de la Rivière:

> The image of the economy which appears from this approach is that of an activity ruled by natural laws, bringing into play physical flows and enduring only through the reproduction of a natural milieu independently of which it could not be analyzed.[57]

Physiocracy's admittedly culturally bound emphasis on the productiveness of agriculture, and of agriculture alone, appears to us now as being unwarranted. But the chief merit of this school remains that it demonstrated a keen understanding of the necessity to preserve the soil, as well as all human resources, in order to ensure the perenity of society.

From then on, economic theory has had more and more to say about less and less. Classical political economy (i.e., A. Smith, D. Ricardo, J. Stuart Mill, J.-B. Say) reduced the reproduction of economic life to that of the market factors;[58] the natural factors, on the contrary, were said to be unvarying and inexhaustible. However, the labor theory of value upon which classical political economy was predicated still implied that the economy was part of a socio-political context from which it could not be separated.[59] This tie, however, was severed by the scarcity theory of value proposed by the neo-classical school (S. Jevons, K. Menger, L. Walras). As Passet explains, i) "the a-temporal equilibrium of the market becomes the central theme of their reflections, to the exclusion of any perspective on reproduction," and ii) "non-market phenomena are rooted out from economics."[60] I concur with Passet that Keynesian macro-economics, despite the appearances, did not reestablish the primacy of societal ends over market factors; besides, it remained silent about environmental issues. While Passet looks at the demise of political economy from an economist's point of view, it is also clear that political theory has been empoverished by this trend. Deprived of its economic dimension, political science can hardly pretend to ever become an applied science and is, therefore, unable to benefit from the kind of practical challenges confronting economists.

On the other hand, Passet aptly remarks that the sphere of economic activities has expanded considerably, to the point where it poses global problems for mankind and the biosphere. Hence the conflict between the logic of natural, global phenomena, and economic rationality, a conflict that is not always clearly analyzed but whose effects are obvious. The solution to this conflict lies, according to Passet, in a "bio-economic" approach. Its goal is to put productive activities back in the natural rhythms of the biosphere and to make them compatible with other spheres of human activities.[61] Unfortunately, Passet, who cogently discusses the methodological implications of this new economic paradigm, has chosen not to examine its political aspects at any great length (even though he did not ignore them). I attempt, hereafter, to complement his analysis in this respect. Indeed, what Passet has done in a regrettably implicit manner, is to pose in practical and often unchallenging terms one of the most pressing political problems of our times, and one that political theory is equally ill-prepared to take up: the problem of the material production of the human habitat. (Of course, stated as such, this awesome problem cannot be circumscribed, let alone resolved within the limits of this section; what follows is meant to be simply an agenda for further research.)

Passet outlines a strategic goal and recommends a series of tactical moves. The goal consists in redefining the finality of the act of production: the pursuit of "*l'avoir*" (possession) must no longer be an end in itself, for man's ultimate concerns are existential ("*l'être*") rather than utilitarian. However, he continues, to *be* truly human we need to live in a social setting. It is impossible to isolate micro-economic rationality from social preferences (and from the contradictions inherent in the formation of the latter). It would like to suggest also that micro-economic models of political life, i.e., the "public choice" approach, which are silent about the socio-political feedbacks built into economic calculus, yield inadequate accounts of the phenomena they purport to explain.

On the other hand, "the primacy of being . . . leads us to favour approaches in terms of reproduction rather than in terms of equilibrium."[62] However, after Quesnay, even the former have excluded the biosphere from their schemata. What Passet does not seem to realize is that this is not so much a theoretical problem for economists as it is a practical political problem which ought to be debated publicly—it is a typical example of what the structuration of the public realm entails in practice.

Passet's tactical prescriptions are well-taken but again fail to take into consideration all the political implications that their implementation would involve. These prescriptions are two-fold, concerning, on the one hand, the method of economic calculus and, on the other hand, institutional design.

In order to bring back the biosphere within the purview of economic theory, it is necessary to look beyond monetary costs and benefits. Energy exchanges and informational contents must also be taken into account, and there exists by now a vast literature suggesting methods for measuring and comparing these factors.[63] There is no need to devote any space here to Passet's analysis of these rather technical questions, especially in view of the fact that I have touched upon similar topics already. But political scientists may wish to pursue this line of research further than Passet cared to take it (or, for that matter, most biologists or economists interested in these questions) by empirically investigating the political factors upon which the implementation of these proposals depends. I see at least two areas of research here: i) the analysis of the attitudes of the bureaucratic and corporate elites towards bio-economic indicators, the degree of awareness of these issues in the attentive public, etc., and ii) the empirical as well as prescriptive investigation of the priority ordering of policy fields where such measures could be introduced by regulation (e.g., agricultural uses of energy vs. transportation, etc.).

We cannot save both nature and our institutions from our own folly, or so it seems to most commentators on the "global predicament" that befalls industrial civilization. Many, including G. Hardin, R. Heilbroner, and W. Ophuls, argue that only some kind of Leviathan could assume the awesome powers required for reversing the mutually reinforcing trends leading us towards the total collapse of civilization. Some authors like E.F. Schumacher, present the case for a massive social and political decentralization.[64] Like S. Beer, and drawing from more or less the same cybernetic and biological sources, Passet declares that this is a false dilemma: centralization is only efficient if it is coupled with decentralization. Global imperatives supersede micro-economic rationality but, at the same time, "every decision ought to be made 'at' and 'by' the organizational level where its consequences are felt."[65]

Admittedly, this terse account oversimplifies Passet's thesis. But I would like to stress the insufficiency of this kind of technical argument (this criticism also applies to Beer's bio-cybernetic model and to other similar positions). What is lacking here is an appreciation of the ethical questions associated with political reforms of this magnitude. Does the "logic of life" invoked by Passet and others point to an ethics of (political) life? No, if by this we mean a ready-made set of normative prescriptions. However, biologists can help us to understand why man is an "ethical animal."[66] Moreover, ecology and ethology can help us to sort out, as it were, the ethical notions that are most relevant to the preservation of life. But we must go even farther. The urgent problem today is not the incompleteness of our scientific models of living systems, but to resist the degradation of the living conditions of perhaps as many as four out of five inhabitants of this planet. In the absence of such an ethical perspective, environmental policy and even Passet's naturalized political economy risk to miss their targets.

Space is lacking here to pursue this line of inquiry much further. But I would like to conclude this discussion with a brief account of E. Morin's thesis according to which fraternity is one of the invariant bases of a universal human nature, the other two being love and intelligence. Apart from the fact that the latter two notions are more immediately understandable (albeit not inherently simpler), they do not find their most direct expression in politics, but rather in religion and science; so that I feel justified in limiting my exegesis of Morin's ideas to his plea for the "fraternalization" of our public lives.

Morin explains that fraternity is something which antedates paternity: while the integration of the father in the family is a crucial step in the hominization of *homo sapiens*, it did not completely supersede the interactive associations among *alter ego* which are characteristic of mammalian societies.[67] It is important to recall that evolution does not erase the past but rearranges it. The hierarchical ordering inherent in the father-child relationship complexifies, but does not cancel what Morin calls the extra-genetic fraternal relationship that binds animal societies together. However, the structuration of the public realm has tended to overvalue the former (e.g., "the fatherland," the father-as-an-authority-figure, etc.) at the expense of the latter. The ethical imperative that follows from the rediscovery of the animality of man consists therefore in a relativization of hierarchical relations among human beings whose individuality is paradoxically grounded in the recognition of *ego* in *alter*. (But, as Morin goes on to note, the genetic, or social, fraternal relation does not only entail a fraternizing potentiality but also a fratricidal one: "it is not sufficient to be brothers to act brotherly" as Romulus and Remus, Cain and Abel remind us!)

Similar concerns have been expressed, albeit from a non-evolutionary point of view, by many authors who lament the atomized lives that people live in modern, urban America (the seminal work in this respect is Riesman's *The Lonely Crowd*).[68] It is also evidently a dominant theme in Critical Theory and in H. Marcuse's work in particular. This somewhat puzzling vision of the one-ness of mankind does not suffice to ground the anti-hierarchical hypothesis built in the biocybernetic models of Passet, Beer, and others, nor does it suffice to legitimize the integration of nation-states in a planetary system. It opens up, however, a new perspective that could eventually lead to such an ethics of life.

To conclude, political theory is an indispensable moment in the making of a public realm. However, it cannot be a blueprint for the latter, for such is the stuff of which ideologies, not authentic theories, are made. Nor can it consist of a set of contextuating "laws of political behavior," for the political animal's radical autonomy is his most essential characteristic. What we are entitled to expect from political theory is an agenda for public debate. Thus I have focused here on one of the most topical, and yet perennially challenging questions of our times: the extent of our own naturalness. Following Morin, I have attempted to show that the most promising way to approach this question is in terms of an "auto-eco-logical" *problématique*. As far as the "autological" (i.e., self-referential) aspects are concerned, we are driven to the conclusion that the political actor's freedom is more firmly grounded in the phenomenology of life than in his ill-fated pursuit of a trans-natural order. The ecological aspects, on the other hand, are less philosophically intriguing but more practically relevant. In this respect, political scientists ought to feel equally—if not more—challenged by Passet's program for a naturalized political economy than economists who are the targeted audience. (In Chapter 6, I have discussed the ecology of collective representations and in Chapter 5 the ecology of action, so to speak.)

Notes

1. The "auto-eco-logical" properties of life are analyzed by Edgar Morin in his *La méthode*, Vol. 2 ("La vie de la vie"), (Paris: Seuil, 1980), pp. 61–69.

2. On the mutually reinforcing relation between the scientific establishment and the political/bureaucratic system see J.J. Salomon, *Science and Politics* (Cambridge, Mass.: MIT Press, 1973).

3. According to S. Moscovici, "the hold that quantic physics or relativist cosmology has upon our history will not turn out to be less profound than the mark left by the French Revolution." (*L'Histoire humaine de la nature*, Paris: Flammarion, 1977, 2nd ed., p. 7.)

4. See the articles by S. Maynard-Moody (on fetal research) and by S. Krimsky (on recombinant DNA research) in Dorothy Nelkin (ed.), *Controversy: Politics of Technical Decisions* (Beverly Hills: Sage, 1979).

5. There exists by now a substantial literature on the relationship between Oriental philosophy and modern science (especially the physics of subatomic particles); one of the most far-reaching accounts of the esoteric concerns evidenced by theoretical physics and biology is to be found in Raymond Ruyer, *La Gnose de Princeton* (Paris: Fayard, 1974); see also F. Capra, *The Tao of Physics* (New York: Bantam Books, 1977). On "Buddhist" economics see E.F. Schumacher, *Small is Beautiful: Economics as if People Mattered* (New York: Harper & Row, 1973).

6. Quoted from *The Crisis of European Sciences and Transcendental Phenomenology* by William Leiss, in *The Domination of Nature* (New York: Braziller, 1972), p. 126. Leiss' book offers an excellent discussion of these questions.

7. R. Iyer, *Parapolitics: Toward the City of Man* (New York: Oxford University Press, 1979), p. 350.

8. See *Communication and the Evolution of Society* (Boston: Beacon Press, 1979), p. 165. Habermas also advances the hypothesis that post-modern societies are confronted again with a demarcation problem but he has in mind another kind of demarcation, namely "the discovery of internal limits which the socialization process runs up against" and "the outbreak of new contingencies at these limits of social individuation," (ibid., p. 167).

9. For Descartes, the body is only a machine whereas the soul is pure thought; by contrast, in Aristotle's view, the soul is manifested in the life of the body.

10. In T.A. Spragens' words, "[t]he disorder of meer nature [Hobbes' phrase to describe an atomized natural world devoid of *telos*], standing as it does in such stark contrast to the orderliness of Aristotelian nature, helps to explain *the radically creative force which Hobbes attributes to the word of the sovereign*. Given the anarchy of meer nature, the voice of the sovereign must serve as the *logos*—the origin of order." *The Politics of Motion* (Lexington: The University of Kentucky Press, 1973), p. 108, my emphasis.

11. According to S. Moscovici, Rousseau severed society from nature insofar as he showed that societal order cannot be construed as a perfectioning of natural order: "After [Rousseau], nature ceased to be the space where by preference one looks for the solutions to the problems befalling men; it neither explained how they had been led to raise them." S. Moscovici, *L'Histoire humaine de la nature* (Paris: Flammarion, 1977), p. 14, my translation.

12. This is a point made by Alvin Toffler in his Foreword to I. Prigogine and I. Stengers, *Order out of Chaos: Man's New Dialogue with Nature* (New York: Bantam Books, 1984).

13. E. Morin, *La méthode* (Paris: Seuil, 1977), vol. 1, p. 274, note that, under Morin's pen, "organization," does not refer to a concrete entity (i.e., a "structure") but to a combination of processes (hence the use of the genetic singular).

14. By "*physis*" Morin means an autopoietic physical reality. Ibid., pp. 17, 274, 276.

15. Ibid., p. 276.

16. A. Giddens, *Central Problems in Social Theory* (London: Macmillan, 1979), p. 71.

17. Ibid., pp. 73–75.

18. Ibid., pp. 73–75. In view of this interesting statement, it is somewhat surprising to find in Giddens' later restatement of his structuration theory a passage in which he paints the difference between the relations of the social and natural sciences to the domain of action in sharp, contrasting tones. I would be more inclined to see there more subtle shades of grey; it is true that social scientific "findings" are more immediately filtered into the world of action, but the practical ramifications of, say, ecology or ethology are not much less evident than those of the sociological literature on suicide (the latter example is borrowed from Giddens' work). See A. Giddens, *The Constitution of Society* (Oxford: Polity Press, 1984), pp. 348–354.

19. For a survey of this new field see Thomas C. Wiegele, *Biopolitics: Search for a More Human Political Science* (Boulder, Co.: Westview Press, 1979).

20. See his introduction to *Biology and Politics* (The Hague: Mouton, 1976).

21. See the papers by G. Schubert, R.D. Masters and D.C. Schwartz in ibid.; see also T.C. Wiegele, *Biopolitics*, pp. 72–79.

22. See the papers by J.C. Davies, B. Tursky, M. Lodge and D. Cross in ibid.; see also G. Shubert, "Psychobiological Politics," *Canadian Journal of Political Science*, XVI: 3 (1983), pp. 535–576.

23. See P.A. Corning, "Toward a Survival Oriented Policy Science," in ibid., pp. 127–154; see also R. Blank, *The Political Implications of Human Genetic Technology* (Boulder, Co.: Westview Press, 1981) and, by the same author, *Redefining Human Life: Reproductive Technologies and Social Policy* (Boulder, Co.: Westview Press, 1984).

24. (See Chapter 4 of this book.) See "Psychobiological Politics."

25. Harvey Sussman, "Neuropsychology and Political Science: The Time is not yet Ripe for Beneficial Interaction," *Politics and the Life Sciences*, August 1985, p. 19.

26. See Douglas Madsen, "A Biochemical Property Relating to Power Seeking in Humans," *American Political Science Review*, vol. 79, no. 2 (1985), pp. 448–457.

27. R.D. Masters, "The Impact of Ethology on Political Science," in A. Somit (ed.), *Biology and Politics*, p. 224.

28. T.C. Wiegele, *Biopolitics*, p. 105.

29. See R.D. Masters, "Evolutionary Biology, Political Theory and the State," *Journal of Social and Biological Structures*, vol. 5 (1982), pp. 439–450. There is also an interesting discussion of the relevance of sociobiology for deciding the question of whether human nature is selfish or social in R. Pettman, *Biopolitics and International Values: Investigating Liberal Norms* (New York: Pergamon, 1981), pp. 42–53; the author concludes that, while there is an innate predisposition toward a kind of self-serving altruism, this tendency has been transformed by symbolic interactions into more complex value systems ranging from communism to classical liberalism.

30. See Northrop Frye, *The Great Code: The Bible and Literature* (Toronto: Academic Press Canada, 1982).

31. See Martin Landau, *Political Theory and Political Science* (New York: Macmillan, 1972), p. 102; Chapter 3 of this text consists in a detailed review of the use and misuse of metaphors in American political thought.

32. *The Great Code*, p. 54.

33. Ibid., p. 15.

34. Ibid., p. 15.

35. Furthermore, within the metonymic and demotic levels, it is also possible to distinguish various degrees of correspondence between the signified and the signifier; for instance, L. von Bertalanffy proposed a scale extending from mere "analogies," i.e., superficial similitudes, to "homologies," i.e., a formal correspondence between the phenomena taken into consideration at the level of explanatory laws, and finally to "isomorphisms," i.e., "structural correspondences" between scientific disciplines. (See L. von Bertalanffy, *General System Theory*, New York: Braziller, 1969, Chapter 3).

36. E.F. Miller, "Metaphor and Political Knowledge," *American Political Science Review*, vol. 73, no. 1 (1979).

37. K. Deutsch's references to concrete cybernetic artifacts are numerous and undisguised; on the other hand, D. Easton's systemic model is unmistakably inspired from a technological version of systems analysis but he has remained silent about the actual origin of his conceptual apparatus.

38. Ibid., pp. 158–160.

39. Ibid., pp. 160–162.

40. Ibid., p. 176.

41. T.A. Spragens, *The Politics of Motion*, p. 177.

42. For a summary of these experiments, see W.H. Thorpe, *Purpose in a World of Chance* (Oxford: Oxford University Press, 1978) pp. 58–73.

43. However, Morin does not ignore the irreducible identity of the phenotype: it is the level at which he discerns the emergence of "knowledge," i.e., the constructive representation of the external world. Thus "animal existence depends, not only on the environment, but on the knowledge of this environment." (*La Méthode*, vol. 2, p. 223).

44. Actually, this "fact" is itself the result of an analogical reasoning linking the societal world to biology!

45. *La Méthode*, vol. 2, p. 190.

46. But let us recall that for Gotthard Günter the very act of computing entails the "logical structure of transjunction," i.e., the ability on the part of the computing entity to reject both terms of an alternative (e.g., true and false) "objectively" describing reality and to substitute for it its own values (see his "Cybernetic Ontology and Transjunctional Operations," in M.C. Yovits *et al.* (eds.) *Self-Organizing Systems*, Washington, D.C.: Spartan Books, 1962, p. 352).

47. *La Méthode*, pp. 159–160.

48. Ibid., p. 185 (my translation).

49. See ibid., pp. 184–186.

50. See ibid., pp. 201–235.

51. See E. Morin, *Le Paradigme perdu: la nature humaine* (Paris: Seuil, 1973).
52. *Central Problems*, p. 5.
53. See John Gray, *Hayek on Liberty* (Oxford: Basil Blackwell, 1984), Chapter 2.
54. *La Méthode*, p. 238.
55. Ibid., pp. 244–254.
56. R.G. Mulgan, "Aristotle's Doctrine that Man is a Political Animal," *Hermes*, vol. 102 (1974), pp. 438–445.
57. *L'Economique et le vivant* (Paris: Payot, 1979), p. 38 (my translation).
58. Ibid., pp. 40–42.
59. Marx and Engels were more aware of the natural implications of capitalist production; nevertheless, they did not include natural factors in their model of the reproduction of capital (ibid., pp. 42–44).
60. Ibid., p. 45.
61. Note that for Passet as well as for Morin (the former cites the latter frequently), the biosphere and the "sphere of human activities" are not completely separated domains; on the contrary, they both partake of the complexity inherent in living systems, and they raise similar methodological problems.
62. Ibid., p. 121.
63. See *inter alia* H.T. Odum, *Energy, Power and Society* (New York: John Wiley, 1971); J. de Rosnay, *The Macroscope* (New York: Harper & Row, 1979), Chapter 3.
64. D.W. Orr and S. Hill, "Leviathan, the Open Society, and the Crisis of Ecology," in D.W. Orr and M.S. Soroos (eds.), *The Global Predicament* (Chapel Hill: The University of North Carolina Press, 1979), pp. 308–326.
65. *L'Economique*, p. 233.
66. See C.H. Waddington, *The Ethical Animal* (Chicago: University of Chicago Press, 1975, reprint).
67. See *La Méthode*, pp. 438–443. Note that the maternal role itself, which is so important in mammalian societies, is the result of evolution; nothing antedates intra-species "fraternal" relations.
68. See, for example, R. Sennet, *The Uses of Disorder* (New York: Vintage Books, 1970), and *The Fall of Public Man* (New York: Alfred A. Knopf, 1977); P.E. Slater, *The Pursuit of Loneliness: American Culture at the Breaking Point* (Boston: Beacon Press, 1970).

Conclusion

Whether or not a social scientific methodology surrenders to the imperatives of technology ought to be the test of its practical relevance. It should be clear at this point that the new cybernetics and the related approaches discussed in this book pass this test. It has been argued that the theoretical relevance of the perspective developed throughout the preceding chapters lies in the opportunity that it offers to political scientists to harmonize their understanding of the scientificity of their discipline with the post-positivist epistemology which is now shared by a wide variety of natural scientists, even if it may not yet qualify as scientific orthodoxy. Complexity and indeterminacy are the foundations of this emerging logic of inquiry. Complexity characterizes the outer appearances of the multidimensional processes through which self-organizing entities develop their domain of autonomy. The phenomenological expression of this autonomy is the identity defining a given complex system; this identity cannot be subsumed in a statistical aggregate and is, therefore, a fundamental source of indeterminacy.

The vision presented in this book is not entirely new. As far as political thought is concerned, several traditions emphasize the complexity and the autonomy of the body politic. What is new here is the possibility of a dialogue between those political scientists who look towards the natural sciences for epistemological guidance and those who insist on the primacy of philosophy over "value-free" science. Another way to put it is to say that the gap between politics and Hobbes' "meer nature" has been bridged, at least in part. To bring about "the triumph of subjectivity in political inquiry," as H.Y. Jung puts it, there is no need to suppose the existence of a radical difference between the goals of science and of philosophy.[1] Existential phenomenologists like H.Y. Jung have not realized that these two cultures are closer to each other than they have been for a very long time. The problem now is to establish the specificity of both politics and political science through critical reflection instead of posing it a priori by *fiat*.

This book has not settled conclusively any specific issue of immediate concern to political scientists, nor was it the intent with which it has been written, unless one is willing to regard the criticisms directed at positivism and scientism as such a demonstration. But it offers a different perspective on a familiar domain, from the standpoint of an evolutionary paradigm of human nature and through a neo-cybernetic model of communication processes. The aspects of the socio-political complex which this perspective highlights are meant to be relevant to the discussion of some of the most pressing questions facing the members of the "post-modern" societies. They include the fundamental importance of crises as moments in the interrelated development of individual and collective identities; the always shifting relations between the regulators and the regulated; the problematic existence of a national psyche; the autonomy of the state as a partial representation of, and a challenge to, the nation; the search for meaning in ordinary political discourse; and the ecological imperative of integrating the biosphere into political economy.

At this point it is appropriate to recapitulate briefly the main points that have been raised in Chapters 5 to 8 regarding the problematic specificity of our subject matter, now that we have come to see even the whole universe as an evolving singularity rather than as the embodiment of timeless and universal "laws." (I shall not restate here the fundamental premises of the new cybernetics and of an evolutionary paradigm of human nature which have been outlined in Chapters 1 to 4.) It is problematic because recursive processes, including politics, always confront us with paradoxes straddling several levels of reality.

As "done-by" we come to experience politics as a constraint on our activities or on the activities of some "doers" from whom we wish to be protected. This situation calls for a cybernetic analysis of the regulatory functions of political authorities. Unfortunately, as it has been argued, the best known systemic and cybernetic models of the political system (e.g., Easton's systems analysis or Deutsch's cybernetic model) do not show clearly how the political system operates as a societal regulator structurally related to most societal subsystems. (We also need more studies of the ways in which a cybernetic analysis can be combined with other approaches to political regulation.) The reproduction of the political system is only one phase in the reproduction of the political community as a whole, a process in which the "done-by" are more personally, if not always more actively, involved than in the operation of the political system. While any autopoietic being possesses an identity whose reproduction is synonymous with the preservation of its life, the symbolic richness of political communications is such that the identification of the identity of a given polity is bound to constitute a contentious issue. I gave the reproduction of a nation-state as an example, not because it is an unambiguous concept, but precisely because it is so vague, and yet so intuitively suggestive.

As civilizations evolve, so do ideas in a symbiotic relationship. A meta-reflection on the self-conscious act of theorizing is needed in order to comprehend both the development of science and the self-organization of advanced capitalism. It appears that the elucidation of the recursive linkages between the natural and social sciences is one way to evoke on a meta-level that cannot be reached, but towards which we can strive, the fundamental unity of life that we have lost sight of. There is a wealth of meaning in the phrase "political life" that I have merely begun to reveal. At the same time, we must think of a way of realizing the unity of political life, of economic life, and of the biosphere. A failure to do so would most certainly result in our alienation from both the human community and the ecological systems upon whose reproduction it depends.

On several previous occasions, suggestions have been offered that could not be taken as far as one might have wished because of space limitations. The question then arises: What steps lay ahead? First, a semiological investigation of the vocabulary in use in both the natural and social sciences is required in order to translate the epistemological loop into actual discourse. This would involve the interpretation of the most important concepts of modern physics and biology, paying particular attention to their socio-political connotations, as well as the relentless pursuit of the naturalistic assumptions hidden in the social scientific terminology. The second and related objective concerns the critical analysis of the many dimensions of societal evolution (e.g., cultural evolution, the evolution of social, economic, and political institutions, etc.); while very suggestive, these analogies could turn out to be misleading if employed too casually. It is imperative to stress the unsettling ambiguity that surrounds the concepts constituent of the theory of evolution. A third direction of research would aim at the development of an open-ended and resolutely normative approach to policy-making designed to render explicit the alternative available to both the "doers" and the "done-by," in the past, present or future. It would be sensitive to the political logic of collective decision-making because it would not be based on optimizing algorithms, but would still be articulated in a manner that would be compatible with the use of formal techniques of computer modeling (e.g., discrete events simulation methods), if such technical refinements are necessary in view of the complexity of the issue at stake.

The final words belong to I. Prigogine. The conclusion of the book that he coauthored with I. Stengers reads:

> We know now that societies are immensely complex systems involving a potentially enormous number of bifurcations exemplified by the variety of

> cultures that have evolved in the relatively short span of human history. We know that such systems are highly sensitive to fluctuations. This leads to both hope and a threat: hope, since even small fluctuations may grow and change the overall structure. As a result, individual activity is not doomed to insignificance. On the other hand, this is also a threat, since in our universe the security of stable, permanent rules seems gone forever.[2]

I hope to have shown that Prigogine's intuitions are indeed well worth pursuing.

Notes

1. See H.Y. Jung, *The Crisis of Political Understanding* (Pittsburgh: Duquesne University Press, 1979), Chapter 9.
2. I. Prigogine and I. Stengers, *Order out of Chaos: Man's New Dialogue with Nature* (New York: Bantam Books, 1984), pp. 312–313.

Glossary

Allopoietic system	A system that is *not* self-producing.
Autopoiesis	The process of self-production and self-renewal in living systems.
Chreod	Developmental path regulating morphological changes.
Chromosome	Rod-shaped bodies into which the genetic material is organized.
Diachronic approach	Focuses on the (irreversible) flow of time through the object of analysis.
Dissipative structures	Self-organizing systems capable of maintaining a state of thermodynamic nonequilibrium, that is, of avoiding entropic decay.
Entropy	The amount of disorganization in a system (measured in units of free energy or information)—but also, under certain conditions, the potentiality of a re-organization.
Epigenesis	Progressive emergence of structural differentiation (starting from an undifferentiated material). In a more restrictive sense, a process of internal selection of genetic information.
Feedback	Input of information about the internal state of a controlled system and/or its relationship with given components of its environment.
"Fuzzyness"	(As in fuzzy set, fuzzy logic, etc.) Uncertainty inherent in sets or classes having imprecise boundaries.
Gene	Unit of information located at a given point on a chromosome and determining a specific hereditary character.

Gene pool	The totality of the hereditary material in a population.
Genotype	The total genetic constitution of an individual organism.
Homeorhesis	Preservation of the stability of a formative process.
Homeostasis	Preservation of a steady state (i.e., a system state in which crucial variables are maintained within an acceptable range of non-life threatening variations).
Morphogenesis	The emergence and evolution of plant or animal forms.
Negentropy	Negative entropy.
Ontogeny	The development of an individual organism.
Organization	The ordering of interactions among individual entities thereby producing a new and complex unity or system.
Othogenesis	Explanation of biological evolution in terms of a finalistic principle.
Phylogeny	The development of a species.
Recursion	A self-referential process.
Self-organization	A process of organizational closure resulting in the production and reproduction of an autonomous system. It is a generic notion that subsumes related and more specific concepts such as autopoiesis or dissipative structure.
Synchronic approach	Focuses on the elements of an object of analysis at a given point in time, or through a period during which entropy is not a significant factor.
Synergism	Combination of interacting elements into a qualitatively new system state.
Variety	The number of possible states that a given system can change into.

Bibliography

Agassi, Joseph. "Methodological Individualism." In John O'Neil, ed. *Modes of Individualism and Collectivism*. New York: St. Martin's, 1972.

Angyal, A. "A Logic of Systems." In F.E. Emery, ed. *Systems Thinking*. Penguin Books, 1969: 17–29.

Ashby, W. Ross. *An Introduction to Cybernetics*. London: Chapman and Hall, 1961.

______. "Setting Goals in Cybernetic Systems." In H.W. Robinson and D.E. Knight, eds. *Cybernetics, Artificial Intelligence and Ecology*. New York: Spartan Books, 1972: 33–44.

Arendt, Hannah. *The Human Condition*. Garden City, New York: Doubleday, 1958.

Atlan, Henri. "Du bruit comme principe d'organisation." *Communications*, vol. 18 (1972): 21–36.

______. *Entre le cristal et la fumée*. Paris: Seuil, 1979.

Attali, Jacques. *La parole et l'outil*. Paris: Presses Universitaires de France, 1975.

______. "L'Ordre par le bruit, le concept de crise en théorie économique." *Communications*, vol. 25 (1976): 86–100.

Barel, Yves. *La reproduction sociale*. Paris: Anthropos, 1973.

______. *Le paradoxe et le système*. Grenoble: Presses Universitaires de Grenoble, 1979.

Bateson, Gregory. *Steps to an Ecology of Mind*. New York: Ballantine, 1972.

Baumgartner, Tom, Walter Buckley, Tom R. Burns, and Peter Schuster. "Meta-Power and the Structuring of Social Hierarchies." In T.R. Burns and W. Buckley, eds. *Power and Control: Social Structures and Their Transformation*. (Beverly Hills: Sage, 1977): 215–288.

Beer, Stafford. *Decision and Control*. New York: John Wiley, 1966.

______. *Brain of the Firm*, 2nd ed. New York: John Wiley & Sons, 1981.

______. *Designing Freedom*. Toronto: CBC Publications, 1974.

______. *Platform for Change*. New York: John Wiley, 1975.

Bell, David V.J. "Methodological Problems in the Study of Political Culture." Paper presented to the 45th Annual Meeting of the Canadian Political Science Association, June 1974.

Bertalanffy, Ludwig von. "The History and Status of General Systems Theory." In G.J. Klir, ed. *Trends in General Systems Theory*. New York: John Wiley, 1972: 21–41.

———. *General System Theory*, 2nd ed. New York: Braziller, 1973.

Birnbaum, Pierre. *La fin du politique*. Paris: Seuil, 1975.

Boulding, Kenneth E. *Beyond Economics: Essays on Society, Religion and Ethics*. Ann Arbor: University of Michigan Press, 1970.

———. *Ecodynamics: A New Theory of Societal Evolution*. Beverly Hills: Sage, 1978.

Brandon, William P. "The Concept of Feedback as an Explanation Teleology." Paper prepared for the 50th Annual Meeting of the Canadian Political Science Association, May 1978.

Brown, G. Spencer. *Laws of Form*. London: George Allen and Unwin, 1969.

Buckley, Walter, ed. *Modern Systems Research for the Behavioral Scientist*. Chicago: Aldine, 1968.

Burns, Tom R. and Walter Buckley, eds. *Power and Control: Social Structures and their Transformation*. Beverly Hills: Sage, 1977.

Cameron, David. *Nationalism, Self-Determination and the Quebec Questions*. Toronto: Macmillan, 1974.

Chodak, Szymon. *Societal Development*. New York: Oxford University Press, 1973.

Coleman, Stephen. *Measurement and Analysis of Political Systems*. New York: John Wiley, 1975.

Coleman, William D. "Political Demands and Political Structures: Some Suggestions Relevant to a System Analytic View of Political Change." Paper presented to the 50th Annual Meeting of the Canadian Political Science Association, May 1978.

Corning, P.S. "Toward a Survival Oriented Policy Science." In A. Somit, ed. *Biology and Politics*. The Hague: Mouton, 1975: 127–154.

———. *The Synergism Hypothesis: A Theory of Progressive Evolution* (New York: McGraw-Hill, 1983).

Crozier, Michael and Erhard Friedberg. *L'Acteur et le système*. Paris: Seuil, 1977.

Davies, J.C. "Ions of Emotion and Political Behavior: A Prototheory." In A. Somit, ed. *Biology and Politics*. The Hague: Mouton, 1975: 97–125.

Deutsch, Karl. *The Nerves of Government*. 2nd ed. New York: Free Press, 1966.

Dobzhansky, Theodosius, Francisco Ayala, J.L. Stebbins, and James W. Valentine. *Evolution*. San Francisco: W.H. Freeman, 1977.

Easton, David. *A Framework for Political Analysis*. Englewood Cliffs, N.J.: Prentice-Hall, 1965.

———. *A Systems Analysis of Political Life*. New York: John Wiley, 1965.

Engels, Friedrich. *Anti Dühring: Herr Eugen Dühring's Revolution in Science*, transl. Emile Burns. New York: International Publishers, 1966.

———. *Dialectics of Nature*. New York: International Publishers, 1960.

Etzioni, Amitai. *The Active Society: A Theory of Societal and Political Processes*. New York: Free Press, 1968.

Eulau, Heinz. *Micro-Macro Political Analysis*. Chicago: Aldine, 1969.

Foerster, Heinz von. "On Self-Organizing Systems and Their Environments." In M. Yovits and S. Cameron eds. *Self-Organizing Systems*. New York: Pergamon Press, 1960: 31–50.

———. "Cybernetics of Cybernetics." In *Cybernetics of Cybernetics*. Urbana, Ill.: Biological Computer Laboratory, University of Illinois, 1974.

______. "Note pour une épistémologie des objets vivants." In E. Morin and M. Piatelli-Palmarini, eds., *L'Unité de l'homme*. Paris: Seuil, 1974: 401–416.

______. "Cybernetics of Cybernetics." *BCL Report*, 13.2 (May 15, 1975).

Forrester, Jay W. *Principles of Systems*. Cambridge, Mass.: Wright-Allen Press, 1968.

Gaudibert, Pierre. "Crise(s) et dialectique." *Communications*, 25 (1976): 118–127.

Geiger, Gebhard. "The Concept of Evolution and Early State Formation." *Politics and the Life Sciences*, 3: 2 (1985): 163–170.

Georgescu-Roegen, Nicholas. *The Entropy Law and the Economic Process*. Cambridge, Mass.: Harvard University Press, 1971.

Giddens, Anthony. *New Rules of Sociological Method*. New York: Basic Books, 1976.

______. *Central Problems in Social Theory*. London: Macmillan, 1979.

______. *The Constitution of Society: Outline of the Theory of Structuration*. Oxford: Polity Press, 1984.

Glucksmann, André. *The Master Thinkers*. New York: Harper and Row, 1980.

Grassé, Pierre-Paul. *Evolution of Living Organisms: Evidence for a New Theory of Transformation*. New York: Academic Press, 1978.

Guillaume, Marc. *Eloge du désordre*. Paris: Gallimard, 1978.

Günther, Gotthard. "Cybernetic Ontology and Transjunctional Operations." In M.C. Yovits *et al*., eds. *Self-Organizing Systems*. Washington, D.C.: Spartan Books, 1962: 313–392.

______. "The Tradition of Logic and the Concept of a Trans-Classic Rationality." *Algemeen Nederlands Tijdschrift woor Wijsbegeerte en Psychologie*. 54E: 4 (1962): 194–200.

Habermas, Jürgen. *Legitimation Crisis*, transl. T. McCarthy. Boston: Beacon Press, 1975.

______. *Communication and the Evolution of Society*, transl. T. McCarthy. Boston: Beacon Press, 1975.

______. *The Theory of Communicative Action*. Vol. I: *Reason and the Rationalization of Society*, transl. T. McCarthy (Boston: Beacon Press, 1984).

Hall, A.D. and R.E. Fagen. "Definition of System." In W. Buckley, ed. *Modern Systems Research for the Behaviorial Scientists*. Chicago: Aldine, 1968: 81–92.

Hines, Samuel M. "Biopolitics and the Evolution of Inquiry in Political Science." *Politics and the Life Sciences*, 1: 1 (1982): 5–16.

Iyer, Raghavan. *Parapolitics: Toward the City of Man*. New York: Oxford University Press, 1979.

Jacob, François. *The Logic of Life: A History of Heredity*. New York: Pantheon Books, 1974.

Jantsch, Erich. *Design for Evolution: Self-Organization and Planning in the Life of Human Systems*. New York: Braziller, 1975.

______, and C.H. Waddington, eds. *Evolution and Consciousness: Human Systems in Transition*. Reading, Mass.: Addison-Wesley, 1976.

______. "Evolution: Self-Realization Through Self-Transcendence." In ibid.: 37–70.

______. *The Self-Organizing Universe*. New York: Pergamon Press, 1980.

Jonas, Hans. *The Phenomenon of Life*. New York: Delta Books, 1969.

Kariel, Henry S. *Open Systems: Arenas for Political Action*. Itasca, New York: Peacock, 1968.

Khoskhish, A. *The Socio-Political Complex: An Interdisciplinary Approach to Political Life*. Oxford: Pergamon Press, 1979.

Koehler, W. "Closed and Open Systems." In F.E. Emery, ed. *Systems Thinking*. Penguin Books, 1969: 59–69.

Koestler, Arthur. *The Ghost in the Machine*. Chicago: Gateway, 1971.

———, and J.R. Smythies, eds. *Beyond Reductionism*. London: Macmillan, 1969.

Laborit, Henri. *La nouvelle grille*. Paris: Laffont, 1974.

Landry, Réjean and Vincent Lemieux. "L'Analyse Cybernétique des politiques gouvernementales." *Canadian Journal of Political Science*. XI (September 1978): 529–544.

Lapierre, Jean-William. *Vivre sans Etat?* Paris: Seuil, 1977.

LaPorte, Todd R., ed. *Organized Social Complexity: Challenge to Politics and Policy*. Princeton, N.J.: Princeton University Press, 1975.

Laszlo, Erwin. *Introduction to Systems Philosophy: Toward a New Paradigm of Contemporary Thought*. New York: Gordon and Breach, 1972.

Leiss, William. *The Domination of Nature*. New York: Braziller, 1972.

Lemieux, Vincent. *Les Cheminements de l'influence*. Quebec: Les Presses de l'Université Laval, 1979.

Lewis, J., ed. *Beyond Chance and Necessity*. London: The Garnstone Press, 1974.

Lumsden, Charles J. and Edward O. Wilson. *Genes, Mind, and Culture: The Coevolutionary Process*. Cambridge, Mass.: Harvard University Press, 1981.

———. *Promethean Fire*. Cambridge, Mass.: Harvard University Press, 1983.

MacKay, Donald M. *Information, Mechanism and Meaning*. Cambridge, Mass.: The MIT Press, 1969.

Madsen, Douglas. "A Biochemical Property Relating to Power Seeking in Humans." *American Political Science Review*. 79: 2 (1985): 448–457.

Maruyama, Magoroh. "The Second Cybernetics: Deviation-Amplifying Mutual Causal Processes." In W. Buckley, ed. *Modern Systems Research for the Behavioral Scientists*. Chicago: Aldine, 1963: 304–313.

Masters, Roger D. "Evolutionary Biology, Political Theory and the State." *Journal of Social and Biological Structures*. 5 (1982): 439–450.

Matson, Floyd W. *The Broken Image: Man, Science and Society*. New York: Braziller, 1964.

Maturana, Humberto R. and Francissco Varela. *Autopoiesis and Cognition: The Realization of the Living*. Boston: D. Reidel, 1980.

McCarthy, Thomas. *The Critical Theory of Jürgen Habermas*. Cambridge, Mass.: The MIT Press, 1978.

Miller, Eugene F. "Positivism, Historism, and Political Inquiry." *American Political Science Review*. (Sept. 1972): 797–817.

———. "Metaphor and Political Knowledge." *American Political Science Review* 73 (March 1979): 155–170.

Miller, G.A. "What is Information Measurement?" In W. Buckley, ed. *Modern Systems Research for the Behavioral Scientists*. Chicago: Aldine, 1968: 123–128.

Miller, James G. *Living Systems*. New York: McGraw-Hill, 1978.

Monod, Jacques. *Chance and Necessity*, transl. A. Wainhouse. New York: Alfred A. Knopf, 1971.

Morin, Edgar. *Le paradigme perdu: la nature humaine*. Paris: Seuil, 1973.

______. "Le retour de l'événement." *Communication*. 18 (1972): 6–20.

______. "Complexity." *International Social Sciences Journal* XXVI: (Dec. 1974): 555–583.

______. "La nature de la société." *Communications*. 22 (1974): 3–32.

______. "Pour une crisologie." *Communications*. 25 (1976): 149–163.

______. *La méthode*. Vol. 1: *La nature de la nature*. Paris: Seuil, 1977. Vol. 2: *La vie de la vie*. Paris: Seuil, 1980.

Moscovici, Serge. *Essai sur l'histoire Humaine de la nature*, 2nd ed. Paris: Flammarion, 1977.

Mulgan, R.G. "Aristotle's Doctrine that Man is a Political Animal." *Hermes*. 102 (1974): 438–445.

Nordlinger, Eric. *On the Autonomy of the Democratic State*. Cambridge, Mass.: Harvard University Press, 1981.

Orr, David W. and Marvin S. Soroos, eds. *The Global Predicament*. Chapel Hill: The University of North Carolina Press, 1979.

______, and Stuart Hill. "Leviathan, the Open Society, and the Crisis of Ecology." In ibid.: 308–326.

Ozbekhan, H. "Toward a General Theory of Planning," in E. Jantsch, ed. *Perspectives of Planning*. Paris: O.E.C.D., 1969: 47–155.

Parsons, Talcott. *The Social System*. New York: Free Press, 1951.

______. "Evolutionary Universals in Society." *American Sociological Review*, 29 (1964): 339–357.

______. *Societies: Evolutionary and Comparative Perspectives*. Englewood Cliffs, N.J.: Prentice-Hall, 1966.

Pask, Gordon. "The Meaning of Cybernetics in the Behavioural Sciences." In J. Rose, ed. *Progress of Cybernetics*, vol. 1. New York: Gordon and Breach, 1969: 15–44.

Passet, René. *L'Economique et le vivant*. Paris: Payot, 1979.

Piaget, Jean. *Biology and Knowledge*, transl. B. Walsh. Chicago: The University of Chicago Press, 1974.

Popper, Karl R. *The Poverty of Historicism*. London: Routledge and Kegan Paul, 1957.

______. *Objective Knowledge*. Oxford: Clarendon Press, 1972.

Poulantzas, Nicos. *Political Power and Social Class*. London: Vergo ed., 1978.

Prigogine, Ilya. "Order through Fluctuations: Self-Organization and Social Systems." In E. Jantsch and C.H. Waddington, eds., *Evolution and Consciousness*. Reading, Mass.: Addison-Wesley, 1976.

______, and I. Stengers. *Order out of Chaos: Man's New Dialogue with Nature*. New York: Bantam Books, 1984.

Rapoport, Anatol. *General System Theory: Essential Concepts and Applications*. Tunbridge Wells, Kent: Abacus Press, 1986.

Roesenblueth, Arturo, Norbert Wiener, and Julian Bigelow. "Behaviour, Purpose and Teleology." In W. Buckley, ed. *Modern Systems Research for the Behavioral Scientists*. Chicago: Aldine, 1968: 221–225.

Rose, S., L.J. Kamin and R.C. Lewontin. *Not in Our Genes: Biology, Ideology and Human Nature*. New York: Pelican Book, 1984.

Rosnay, Joel de. *The Macroscope*. New York: Harper and Row, 1979.

Schroedinger, E. *What is Life?* New York: Macmillan, 1947.

Service, E.R. *Origins of the State and Civilization: The Process of Cultural Evolution*. New York: Norton, 1975.

Schubert, Glendon. "Psychobiological Politics." *Canadian Journal of Political Science*. XVI: 3 (1983): 535–576.

Simon, Herbert A. *The Sciences of the Artificial*. Cambridge, Mass.: The MIT Press, 1969.

Somit, Albert. *Biology and Politics: Recent Explorations*. Paris: Mouton, 1976.

Spencer, Herbert. *On Social Evolution*, ed. J.D.Y. Peel. Chicago: The University of Chicago Press, 1972.

Starling, J. "The Use of Systems Constructs in Simplifying Organized Social Complexity." In T.D. LaPorte, ed. *Organized Social Complexity*. Princeton, N.J.: Princeton University Press, 1975: 151–172.

Steinbruner, John. *Cybernetic Theory of Decision: New Dimensions of Political Analysis*. Princeton, N.J.: Princeton University Press, 1974.

Sutherland, John W. *Societal Systems: Methods, Modeling and Management*. New York: Elsevier North-Holland, 1978.

Thom, René. *Structural Stability and Morphogenesis: An Outline of a General Theory of Models*. Reading, Mass.: Benjamin Cummings, 1975.

Thorson, Thomas L. *Biopolitics*. New York: Holt, Rinehart and Winston, 1971.

Touraine, Alain. *The Self-Production of Society*, transl. D. Coltman. Chicago: The University of Chicago Press, 1977.

Varela, Francisco G. "The Grounds for a Closed Logic." *BCL Report*, no. 35 (1975).

______. *Principles of Biological Autonomy*. New York: Elsevier North-Holland, 1979.

______, Humberto R. Maturana, and R. Uribe. "Autopoieisis: the Organization of Living Systems, its Characterization and a Model." *Bio-systems* 5: 4 (1974): 187–196.

Vickers, Geoffrey. *Freedom in a Rocking Boat: Changing Values in an Unstable Society*. Penguin Books, 1970.

______. "Values, Norms and Policies." *Policy Sciences* 4 (1973): 103–111.

Waddington, Conrad H., ed., *Towards a Theoretical Biology*, 4 vols. Edinburgh: Edinburgh University Press, 1969–1971.

______. *The Evolution of an Evolutionist*. Edinburgh: Edinburgh University Press, 1975.

______. *Tools for Thought: About Complex Systems*. New York: Basic Books, 1978.

Weaver, Warren. "Science and Complexity." *American Scientist*. 36 (1948): 536–566.

Whitehead, Alfred North. *Science and the Modern World*. New York: Free Press, 1967.

Wiegele, Thomas C. *Biopolitics: Search for a More Human Political Science*. Boulder, Co.: Westview Press, 1979.

Wiener, Norbert. *Cybernetics or Control and Communication in the Animal and the Machine*, 2nd ed. Cambridge, Mass.: The MIT Press, 1961.

Wilson, Edward O. *Sociobiology: The New Synthesis.* Cambridge, Mass.: Harvard University Press, 1975.

Wolin, Sheldon. *Politics and Vision: Continuity and Innovation in Western Political Thought.* Boston: Little, Brown, 1960.

Wright, Georg H. von. *Explanation and Understanding.* London: Routledge and Kegan Paul, 1971.

Yovits, M.C. and S. Cameron, eds. *Self-Organizing Systems.* New York: Pergamon Press, 1960.

———, G.I. Jacobi, and G.D. Goldstein, eds. *Self-Organizing Systems.* Washington, D.C.: Spartan Books, 1962.

Zeleny, Milan, ed. *Autopoiesis, Dissipative Structures, and Spontaneous Social Orders.* Boulder: Westview, 1980.

———, ed. *Autopoiesis.* New York: North Holland, 1981.

Index

For Product Safety Concerns and Information please contact our EU representative GPSR@taylorandfrancis.com Taylor & Francis Verlag GmbH, Kaufingerstraße 24, 80331 München, Germany

Batch number: 10725915

Printed by Printforce, the Netherlands